AF479088

Series Editor
John M. Walker
School of Life Sciences
University of Hertfordshire
Hatfield, Hertfordshire, AL10 9AB, UK

For further volumes:
http://www.springer.com/series/7651

Engineering and Analyzing Multicellular Systems

Methods and Protocols

Edited by

Lianhong Sun

*School of Life Sciences, University of Science & Technology of China,
Hefei, Anhui, People's Republic of China*

Wenying Shou

Division of Basic Sciences, Fred Hutchinson Cancer Research Center, Seattle, WA, USA

Humana Press

Editors
Lianhong Sun
School of Life Sciences
University of Science & Technology of China
Hefei, Anhui, People's Republic of China

Wenying Shou
Division of Basic Sciences
Fred Hutchinson Cancer Research Center
Seattle, WA, USA

ISSN 1064-3745 ISSN 1940-6029 (electronic)
ISBN 978-1-4939-0553-9 ISBN 978-1-4939-0554-6 (eBook)
DOI 10.1007/978-1-4939-0554-6
Springer New York Heidelberg Dordrecht London

Library of Congress Control Number: 2014934331

Humana Press is a brand of Springer
Springer is part of Springer Science+Business Media (www.springer.com)

Preface

Microbial ecosystems consist of many interacting microbial species. With synthetic biology rapidly evolving from engineering genetic circuits within cells to manipulating cell–cell interactions, several synthetic microbial communities have been constructed. Such synthetic communities have been used in basic research to explore questions such as how interactions within a community shape the stability, function, patterning, and evolution of the community. In addition, synthetic communities have been constructed to solve challenging engineering problems in various fields. As a consequence, a framework of engineering synthetic microbial ecosystems/consortia is of importance to many users. Equally important are the transcriptomic, genomic, cell biological, and chemical methods to characterize communities. Ultimately, to quantitatively understand complex microbial communities, predictive mathematical models can be extremely useful. This volume of Methods in Molecular Biology includes recent developments and a variety of examples on how to construct, analyze, and mathematically model multicellular systems.

Hefei, People's Republic of China *Lianhong Sun*
Seattle, WA, USA *Wenying Shou*

Contents

Contents

Contributors

SHOTARO AYUKAWA • *Academy of Computational Life Sciences, Tokyo Institute of Technology, Kanagawa, Japan*

JEFFREY E. BARRICK • *Department of Molecular Biosciences, Center for Systems and Synthetic Biology, Center for Computational Biology and Bioinformatics, Institute for Cellular and Molecular Biology, The University of Texas at Austin, Austin, TX, USA*

HANS C. BERNSTEIN • *Department of Chemical and Biological Engineering, Center for Biofilm Engineering, Montana State University, Bozeman, MT, USA*

KRISTEN A. BRILEYA • *Department of Microbiology, Center for Biofilm Engineering, Montana State University, Bozeman, MT, USA*

LAURA B. CAMILLERI • *Department of Microbiology, Center for Biofilm Engineering, Montana State University, Bozeman, MT, USA*

ROSS P. CARLSON • *Department of Chemical and Biological Engineering, Center for Biofilm Engineering, Montana State University, Bozeman, MT, USA*

MATTHEW WOOK CHANG • *Department of Biochemistry, Yong Loo Lin School of Medicine, National University of Singapore, Singapore, Singapore*

LON CHUBIZ • *Department of Organismic and Evolutionary Biology, Harvard University, Cambridge, MA, USA*

GEORGE M. CHURCH • *Department of Genetics, Harvard Medical School, Boston, MA, USA; Wyss Institute for Biologically Inspired Engineering, Harvard University, Boston, MA, USA*

DANIEL E. DEATHERAGE • *Department of Molecular Biosciences, Center for Systems and Synthetic Biology, Center for Computational Biology and Bioinformatics, Institute for Cellular and Molecular Biology, The University of Texas at Austin, Austin, TX, USA*

SARAH DOUGLAS • *Department of Organismic and Evolutionary Biology, Harvard University, Cambridge, MA, USA*

MATTHEW W. FIELDS • *Department of Microbiology, Center for Biofilm Engineering, Montana State University, Bozeman, MT, USA*

XIONGFEI FU • *Department of Physics, The University of Hong Kong, Pokfulam, Hong Kong, China*

MARTIN FUSSENEGGER • *Department of Biosystems Science and Engineering, ETH Zurich, Basel, Switzerland*

ROBIN GREEN • *Molecular and Cellular Biology Program, University of Washington, Seattle, WA, USA; Division of Basic Sciences, Fred Hutchinson Cancer Research Center, Seattle, WA, USA*

WILLIAM HARCOMBE • *Department of Ecology, Evolution, and Behavior, University of Minnesota, St. Paul, MN, USA*

BRANDI S. HEATH • *Physical Sciences Division, Pacific Northwest National Laboratory, Richland, WA, USA*

BOON CHIN HENG • *Department of Biosystems Science and Engineering, ETH Zurich, Basel, Switzerland*

KRISTINA HILLESLAND • *Biological Sciences Division, School of STEM, UW Bothell, Bothell, WA, USA*

KAZUFUMI HOSODA • *Graduate School of Information Science and Technology, Osaka University, Suita, Osaka, Japan*

WEI HUANG • *Department of Biology, South University of Science and Technology of China Nanshan, Shenzhen, China; Department of Biochemistry, The University of Hong Kong, Pokfulam, Hong Kong, China*

IN YOUNG HWANG • *Department of Biochemistry, Yong Loo Lin School of Medicine, National University of Singapore, Singapore, Singapore*

SIMON KAHAN • *Northwest Institute for Advanced Computing, University of Washington, Seattle, WA, USA*

SEUNGHWA KANG • *Pacific Northwest National Laboratory, Seattle, WA, USA*

DAISUKE KIGA • *Department of Computational Intelligence and Systems Science, Tokyo Institute of Technology, Kanagawa, Japan*

JULIA LASKIN • *Physical Sciences Division, Pacific Northwest National Laboratory, Richland, WA, USA*

SUJUNG LIM • *Biological Sciences Division, School of STEM, UW Bothell, Bothell, WA, USA*

HUA LING • *Department of Biochemistry, Yong Loo Lin School of Medicine, National University of Singapore, Singapore, Singapore*

CHUEH LOO POH • *School of Chemical and Biomedical Engineering, Nanyang Technological University, Singapore, Singapore*

MATTHEW J. MARSHALL • *Biological Sciences Division, Pacific Northwest National Laboratory, Richland, WA, USA*

BABAK MOMENI • *Division of Basic Sciences, Fred Hutchinson Cancer Research Center, Seattle, WA, USA*

NAOAKI ONO • *Graduate School of Information Science, Nara Institute of Science and Technology, Ikoma, Nara, Japan*

BAHAREH HAJI RASOULIHA • *School of Chemical and Biomedical Engineering, Nanyang Technological University, Singapore, Singapore*

RYOJI SEKINE • *Department of Computational Intelligence and Systems Science, Tokyo Institute of Technology, Kanagawa, Japan*

WENYING SHOU • *Division of Basic Sciences, Fred Hutchinson Cancer Research Center, Seattle, WA, USA*

SERGEY STOLYAR • *Institute for Systems Biology, Seattle, WA, USA*

SHINGO SUZUKI • *RIKEN Quantitative Biology Center, Furuedai, Osaka, Japan*

MUI HUA TAN • *School of Chemical and Biomedical Engineering, Nanyang Technological University, Singapore, Singapore*

ADAM JAMES WAITE • *Molecular and Cellular Biology Program, University of Washington, Seattle, WA, USA; Division of Basic Sciences, Fred Hutchinson Cancer Research Center, Seattle, WA, USA*

HARRIS H. WANG • *Department of Systems Biology, Columbia University Medical Center, New York, NY, USA; Department of Pathology and Cell Biology, Columbia University Medical Center, New York, NY, USA*

CHOON KIT WONG • *School of Chemical and Biomedical Engineering, Nanyang Technological University, Singapore, Singapore*

STEPHANIE J. YAUNG • *Program in Medical Engineering Medical Physics, Harvard-MIT Health Sciences and Technology, Cambridge, MA, USA; Department of Genetics, Harvard Medical School, Boston, MA, USA; Wyss Institute for Biologically Inspired Engineering, Harvard University, Boston, MA, USA*

TETSUYA YOMO • *Graduate School of Information Science and Technology, Osaka University, Suita, Osaka, Japan; Graduate School of Frontier Bioscience, Osaka University, Suita, Osaka, Japan; Exploratory Research for Advanced Technology, Suita, Osaka, Japan*

CHUN YOU • *Biological Systems Engineering Department, Virginia Tech, Blacksburg, VA, USA*

XIAO-ZHOU ZHANG • *Biological Systems Engineering Department, Virginia Tech, Blacksburg, VA, USA; Gate Fuels Inc., Blacksburg, VA, USA*

YI-HENG PERCIVAL ZHANG • *Biological Systems Engineering Department, Virginia Tech, Blacksburg, VA, USA; Gate Fuels Inc., Blacksburg, VA, USA*

Part I

Constructing Multicellular Systems

Chapter 1

Recent Progress in Engineering Human-Associated Microbiomes

Stephanie J. Yaung, George M. Church, and Harris H. Wang

Abstract

Recent progress in molecular biology and genetics opens up the possibility of engineering a variety of biological systems, from single-cellular to multicellular organisms. The consortia of microbes that reside on the human body, the human-associated microbiota, are particularly interesting as targets for forward engineering and manipulation due to their relevance in health and disease. New technologies in analysis and perturbation of the human microbiota will lead to better diagnostic and therapeutic strategies against diseases of microbial origin or pathogenesis. Here, we discuss recent advances that are bringing us closer to realizing the true potential of an engineered human-associated microbial community.

Key words Microbiome, Microbiota, Synthetic biology, Systems biology, Microbial engineering, Functional metagenomics, Host–microbe interactions

1 Introduction

Of the 100 trillion cells in the human body, 90 % are microbes that naturally inhabit various body sites, including the gastrointestinal tract, nasal and oral cavities, urogenital area, and skin [1]. An individual's colon is home to 10^{11}–10^{12} microbial cells/mL, the greatest density compared to any other microbial habitat characterized to date [2]. Many studies, such as the Human Microbiome Project and MetaHIT, have probed the vast effects of microbiota on human health and disease [1, 3–5]. In addition to metagenomic sequencing [6], traditional methods of studying cells in isolation are important for elucidating molecular bases of microbial activity. However, cells do not exist in single-species cultures in nature. In fact, some species are only culturable in the presence of other microorganisms [7]. This interdependence for survival amongst microbial species in a community attests to the importance of intercellular interactions, both microbe–microbe and host–microbe. Despite the fact that the human microbiota is composed of many individual microbes, these individuals work in concert to

Lianhong Sun and Wenying Shou (eds.), *Engineering and Analyzing Multicellular Systems: Methods and Protocols,*
Methods in Molecular Biology, vol. 1151, DOI 10.1007/978-1-4939-0554-6_1, © Springer Science+Business Media New York 2014

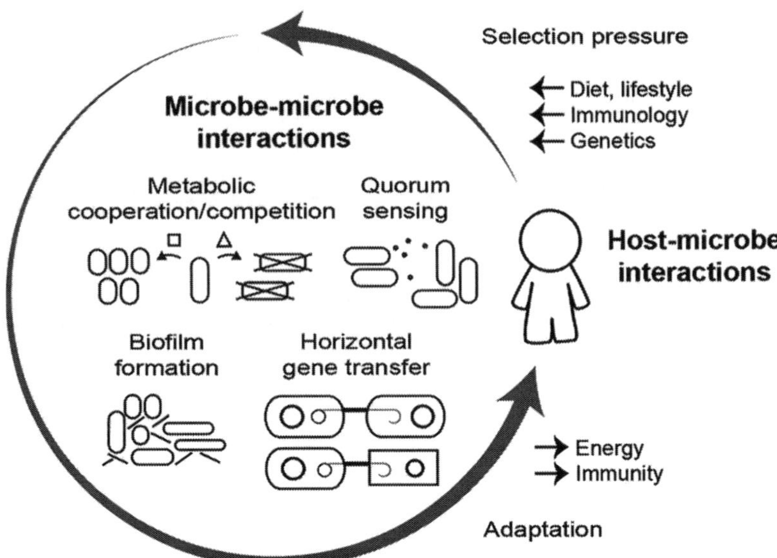

Fig. 1 Engineering human-associated microbiota requires detailed understanding of processes that govern the natural propagation and retention of microbes in the host as well as environmental and adaptive pressures that drive the evolution of cells and communities

perform tasks that rival in complexity to those of more sophisticated multicellular systems. Thus, the human-associated microbiome presents a ripe opportunity for forward engineering to potentially improve human health (Fig. 1). Here, we review recent advances in this area and outline potential avenues for future endeavors.

2   Microbiota, Host, and Disease

Contrary to traditional views, microbes are social organisms that engage with the environment and other organisms in specific ways. Microbes participate in intercellular communication through contact-dependent signaling [8], quorum sensing [9], metabolic cooperation or competition [5], spatiotemporal organization [10], and horizontal gene transfer (HGT) [11]. Human-associated microbes produce by-products that serve as substrates utilized by other resident bacteria [12–14]. For instance, accumulated hydrogen gas from bacterial sugar fermentation is removed by acetogenic, methanogenic, and sulfate-reducing gut bacteria [15]. In contrast to cross-feeding relationships, microbes under stress can release bacteriocins to suppress the growth of competitors [16–18]. If microbes are members of a biofilm community, they benefit from physical protection from the environment, access to nutrients trapped and distributed through channels in the biofilm, development of syntrophic relationships with other members, and the ability to share and acquire genetic traits [19, 20]. Microbial populations also

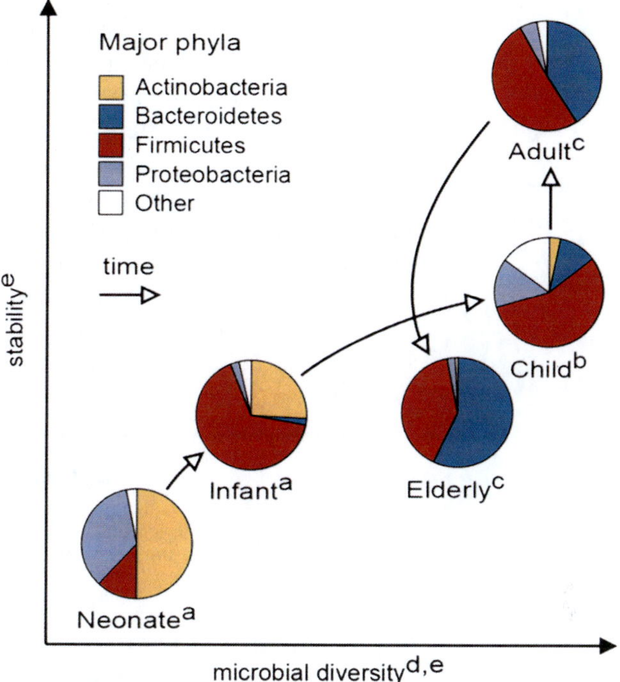

Fig. 2 Composition of the human gut microbiome during development with respect to microbial diversity and population stability. Data compiled from recent studies from the literature: (*a*) Hong 2010 [169]; (*b*) Saulnier 2011 [170]; (*c*) Claesson 2011 [171]; (*d*) Yatsunenko 2012 [172]; (*e*) Spor 2011 [173]

genetically diversify to insure against possible unstable environmental conditions [21, 22]. Moreover, multispecies communities harbor a dynamic gene pool consisting of mobile genetic elements, such as transposons, plasmids, and bacteriophages, which serve as a source of HGT to share beneficial functions with neighbors to preserve community stability [23–26]. Densely populated communities such as the human gut are active sites for gene transfer and reservoirs for antibiotic resistance genes [11, 27–29].

Beyond microbe–microbe interactions, the microbiota coevolves with the host as it develops, driving microbial adaptation [30–33]. Core functions of microbiota benefit the host, such as extraction of otherwise inaccessible nutrients, immune system development, and protection against pathogen colonization [2, 34–37]. Gut microbes are critical in intestinal angiogenesis, epithelial cell maturation, and immunological homeostasis [37–40]. For example, the commensal *Bacteroides fragilis* produces polysaccharide A, which converts host CD4[+] T cells into Foxp3[+] T_{reg} cells, producing interleukin-10 (IL-10) and inducing mucosal tolerance [41]. Host diet, inflammatory responses, and aging also affect microbial community composition and function [42–45] (Fig. 2). Indeed, aberrations in host genetics, immunology, and diet can lead to microbiota-associated human diseases. Diet-induced obesity in mice from a high-fat diet is characterized by enhanced energy harvest and an increased *Firmicutes*-to-*Bacteroidetes* ratio [46, 47]. Furthermore, disruptions in the homeostasis between gut microbial antigens and host immunity can invoke allergy and autoimmunity, as in type 1 diabetes and multiple sclerosis [48–50]. It is thought that inflammatory bowel disease (IBD) results from inappropriate immune responses to intestinal bacteria; genes identified in genome-wide association studies highlight the role of a host imbalance between pro-inflammatory and regulatory states [48, 51].

While the host selects for microbial communities that harvest nutrients and prime the immune system, irregular microbiota composition may cause disease (Fig. 3), including IBD [52–54],

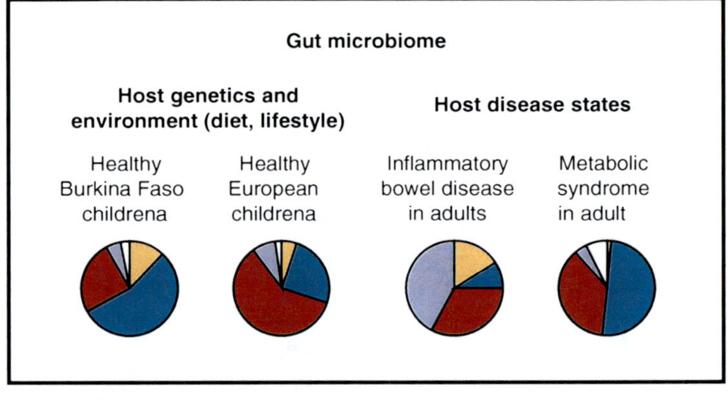

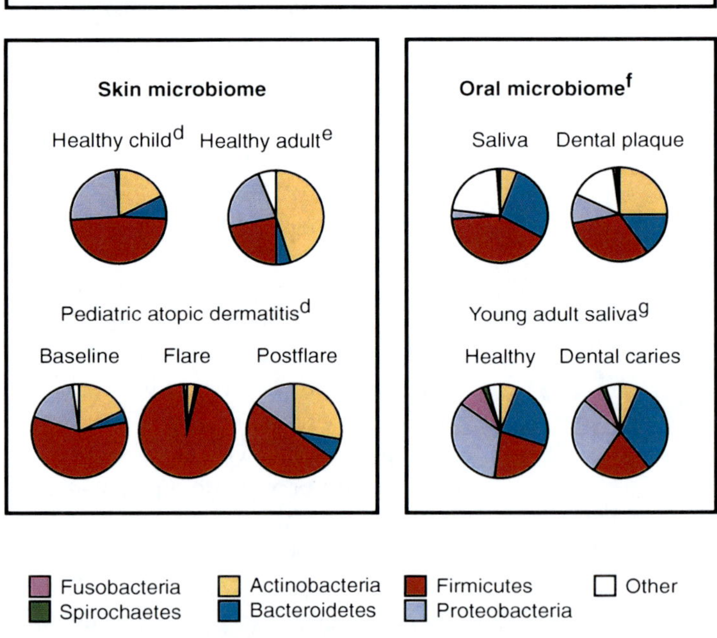

Fig. 3 Changes in the composition of human microbiota during disease states compared to healthy states. Data compiled from recent studies from the literature: (*a*) De Filippo 2010 [174]; (*b*) Peterson 2008 [175]; (*c*) Larsen 2010 [176]; (*d*) Kong 2012 [177]; (*e*) Gao 2012 [178]; (*f*) Keijser 2008 [179]; (*g*) Yang 2012 [180]

lactose intolerance [55, 56], obesity [57, 58], type I diabetes [59], arthritis [60], myocardial infarction severity [61], and opportunistic infections by pathogens such as *Clostridium difficile* and HIV [62–65]. Microbial gut metabolism links host diet not only to body composition and obesity [66] but also to chronic inflammatory states, such as IBD, type 2 diabetes, and cardiovascular disease [67–69]. Intestinal microbes are also important in off-target drug metabolism, rendering digoxin, acetaminophen, and irinotecan less effective or even toxic [70–72]. In the case of irinotecan, a chemotherapeutic used mainly for colon cancer, the drug is metabolized by β-glucuronidases of commensal gut bacteria into a toxic form that damages the intestinal lining and causes severe diarrhea. In the oral cavity, ecological shifts in dental plaque microbiota lead to caries (cavities), gingivitis, and periodontitis [73]. Dental caries arise from acidic environments generated by acidogenic (acid-forming) and aciduric (acid-tolerant) bacteria, which metabolize sugar from the host diet. Translocation of oral bacteria into other tissues results in infections, and cytokines from inflamed gums released into the bloodstream stimulate systemic inflammation. Oral bacteria have been implicated in respiratory [74, 75] and cardiovascular diseases [76–78], though mechanisms remain unclear.

3 Enabling Tools for Engineering the Microbiota

The human-associated microbial community presents a vast reservoir of nonmammalian genetic information that encodes for a variety of functions essential to the mammalian host [79]. Next-generation sequencing technologies have enabled us for the first time to systematically probe the genetic composition of these trillions of microbes that reside on the human body [1]. The ongoing effort by the Human Microbiome Project and MetaHIT to catalog dominant microbial strains from different body sites has generated useful reference genomes for many of the representative species [80]. Metagenomic shot-gun sequencing approaches of whole microbial communities, such as those found in the gut, have yielded near-complete gene catalogs that describe abundance and diversity of genes that contribute to maintenance and metabolism of the microbiota [6].

In order to determine functional relationships between human-associated microbes and their concerted effect in the mammalian host, we rely on functional perturbation of the microbial community. These investigative avenues include genome-scale perturbation assays, specified community reconstitutions, and directed engineering through synthetic biology (Fig. 4). Each approach provides us with a unique angle to attack an otherwise daunting

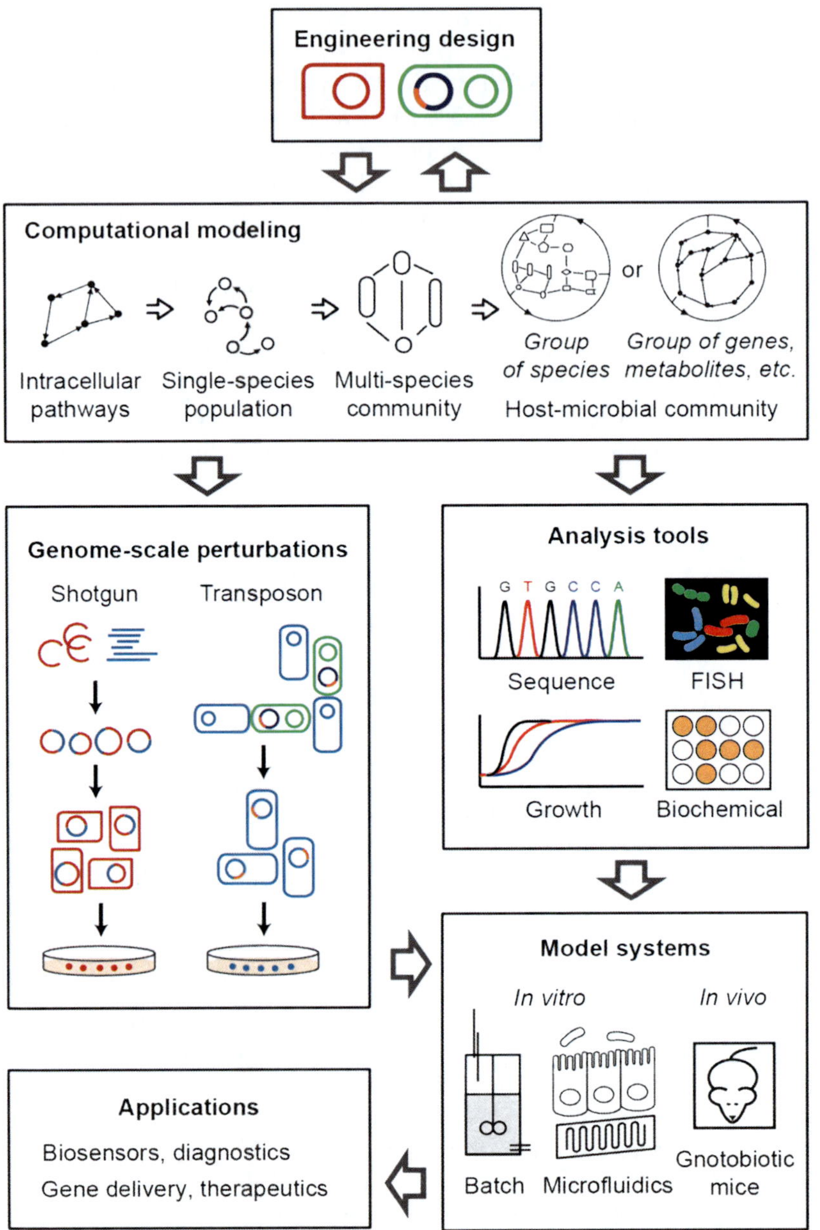

Fig. 4 General approaches to engineer the human microbiome through design, quantitative modeling, genome-scale perturbation, and analysis in in vitro and in vivo models, with the ultimate goal of producing demand-meeting applications to improve sensing, prevention, and treatment of diseases

challenge of de-convolving a highly intertwined set of microbial interactions in a very heterogeneous environment and a difficult-to-manipulate human host. Advances in both in vitro and in vivo host models have thus also facilitated research endeavors in this area, which we discuss in the following sections.

3.1 Challenges of Building New Genetic System

Approaches to study the function of human-associated microbes by genetic manipulation rely on several fundamental capabilities, which are often the largest practical barriers to manipulate microbes genetically. First, individual microbes need to be isolated and cultured in the laboratory. Because microbes have a myriad of physiologies and require different nutritional supplement for growth, different media compositions and growth conditions need to be laboriously tested by trial and error to isolate and culture each microbe. These microbial culturing techniques date back to the times of Louis Pasteur and are still the dominant approach today. More recent microbial cultivation techniques use microfluidics and droplet technologies to enable the discovery of synergistic interactions between natural microbes that allow otherwise "unculturable" organisms to be grown in laboratory conditions [7, 81, 82].

Upon successful microbial cultivation, the next limiting step of microbial genetic manipulation is the transformation of foreign DNA into cells. The passage of foreign DNA (e.g., plasmids, recombinant fragments) into the cell requires overcoming the physical barriers presented by the cell wall or membrane. This task is accomplished in nature through processes such as transduction by phage, conjugation and mating, or natural competency and DNA uptake [83, 84]. Numerous laboratory techniques have been developed for microbial transformation including electroporation [85], biolistics [86], sonication [87], and chemical or heat disruption [88]. Electroporation, the most common of the laboratory transformation techniques, relies on high-voltage electrocution of the bacterial sample that is thought to transiently induce pores on the cell membrane (hence "electroporation") that then enable extracellular DNA to diffuse into the cell. Various protocols for electroporation of human-associated microbes have been described and are good starting points for developing genetic systems in these microbes [89, 90].

Upon transformation of DNA into the cell, the DNA needs to either stably propagate intracellularly or integrate into the microbial host genome through recombination or other integration strategies. Inside the cell, stable propagation of episomal DNA such as plasmids requires DNA replication machinery that is compatible with the foreign DNA [83]. Additionally, cells often use methylation and DNA modification and restriction systems to discern foreign versus host DNA through a primitive defensive mechanism that fights against viruses or other invading genetic elements. Nonetheless, these promiscuous genetic elements can often be used as a way to integrate foreign DNA into the chromosome and are often used for large-scale functional genomics [91].

Taking all these parameters into consideration, we have summarized (Fig. 5) the current genetic tractability of human-associated microbes with respect to culturability, availability of full genome sequences, transfection methods, and expression and manipulation systems. Expansion of these basic genetic tools is crucial for future functional studies of human microbiota.

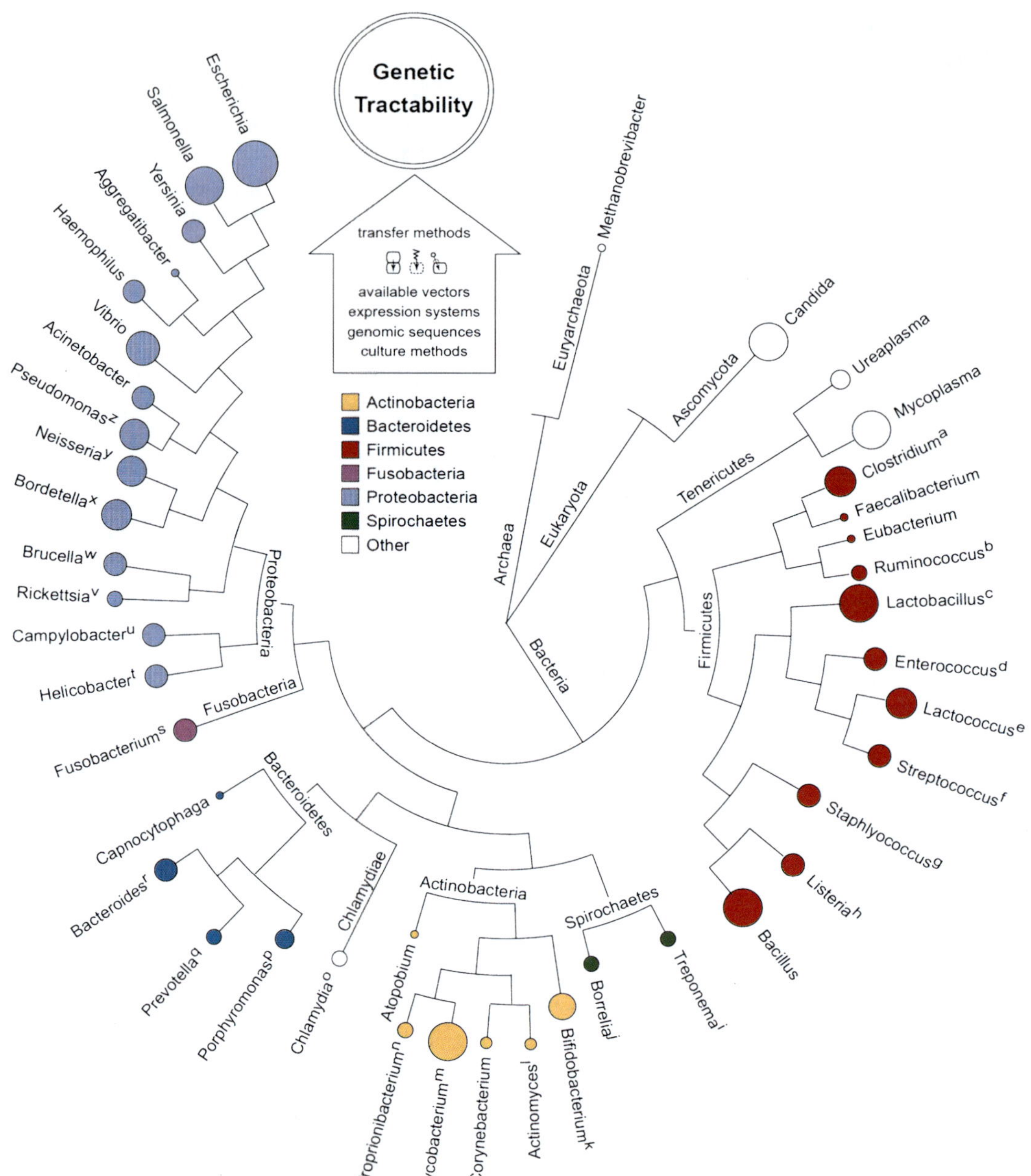

Fig. 5 Genetic tractability of abundant or relevant human-associated microbial genera, evaluated by the availability of means to introduce genetic material (e.g., transformation, conjugation, or transduction), vectors, expression systems, completed genomic sequences, and culturing methods. *Circles* of increasing sizes indicate greater genetic tractability. Protocols and demonstrated methods for genetic manipulation are listed as follows: (*a*) Clostridium: Phillips-Jones 1995, Jennert 2000, Young 1999, Bouillaut 2011 [181–184]; (*b*) Ruminococcus: Cocconcelli 1992 [185]; (*c*) Lactobacillus: van Pijkeren 2012, Ljungh 2009, Damelin 2010, Sorvig 2005, Thompson 1996, Lizier 2010[107, 186–190]; (*d*) Enterococcus: Shepard 1995 [191]; (*e*) Lactococcus: Holo 1995, van Pijkeren 2012 [107, 192]; (*f*) Streptococcus: McLaughlin 1995, Biswas 2008 [193, 194]; (*g*) Staphlyococcus: Lee 1995 [195]; (*h*) Listeria: Alexander 1990 [196]; (*i*) Treponema: Kuramitsu 2005 [197]; (*j*) Borrelia: Hyde 2011, Rosa 1999 [198, 199]; (*k*) Bifidobacterium: Mayo 2010 [200];

3.2 Genome-Scale Perturbations

Genome-scale perturbations are a class of genetic approaches that disrupt or perturb the expression of functional genes that contribute to relevant phenotypes by individual microbes. To dissect the function of different genes in the cell, we have relied heavily on the use of transposons, which are selfish genetic elements that can splice into and out of different locations of chromosomal DNA, thereby disrupting the coding sequence [92]. This classical approach, known as transposon mutagenesis, has allowed us to isolate many genetic mutants whose disrupted genes give rise to interesting phenotypes that reflect the importance of those genes to its physiology. Next-generation DNA sequencing has now enabled multiplexed genotyping of pools of transposon mutants by using molecular barcodes that then can be applied to measure the effect of genome-scale perturbations in different environmental conditions. For example, techniques such as insertion sequencing (INSeq) [93] utilize the inverted repeat recognition of the Himar transposase, which is one nucleotide change away from the restriction site for type II restriction enzyme MmeI, to generate paired 16–17 bp flanking genomic sequences around the transposon that can be sequenced in pools. Thus, the defined insertion location of every transposon in the library can be determined. By sequencing this pooled mutant library pre- and posttreatment with any number of environmental perturbations, one can probe the effects of different gene disruptions on the physiology of the cell in a multiplexed fashion. Similar techniques using other transposon systems such as transposon sequencing (Tn-seq) [94], high-throughput insertion tracking by deep sequencing (HITS) [95], and transposon-directed insertion-site sequencing (TraDIS) [96] have also been developed.

In addition to transposon-based systems, shotgun expression libraries have been useful in discovering functional DNA elements in genomic or metagenomic DNA. Shotgun expression libraries rely on physical shearing or restriction digestion of a donor DNA source into smaller DNA fragments that are then cloned into a gene expression vector and transformed into a host strain for functional analysis. A library of metagenomic DNA samples can for example be extracted from an environment and cloned into plasmids that are then expressed in *E. coli*. Selection and sequencing of the *E. coli* population for heterologous DNA that enable new function lead to discovery of novel gene elements that perform a particular

---

Fig. 5 (continued) (*l*) Actinomyces: Yeung 1994 [201]; (*m*) Mycobacterium: Parish 2009, Sassetti 2001 [202, 203]; (*n*) Proprionibacterium: Luijk 2002 [204]; (*o*) Chlamydia: Binet 2009 [205]; (*p*) Porphyromonas: Belanger 2007 [206]; (*q*) Prevotella: Flint 2000, Salyers 1992 [207, 208]; (*r*) Bacteroides: Salyers 1999, Smith 1995, Bacic 2008 [209–211]; (*s*) Fusobacterium: Haake 2006 [212]; (*t*) Helicobacter: Taylor 1992, Segal 1995 [213, 214]; (*u*) Camplyobacter: Taylor 1992 [214]; (*v*) Rickettsia: Rachek 2000 [215]; (*w*) Brucella: McQuiston 1995 [216]; (*x*) Bordetella: Scarlato 1996 [217]; (*y*) Neisseria: O'Dwyer 2005, Bogdon 2002, Genco 1984 [218–220]; (*z*) Pseudomonas: Dennis 1995 [221]

activity. This approach can easily identify activities such as antibiotic resistance [97] but have yielded less success with other functions.

Towards forward engineering of human-associated microbes, new genome engineering tools such as trackable multiplex recombineering (TRMR) [98, 99] and multiplex automated genome engineering (MAGE) enable efficient, site-specific modification of the genome [100–103]. TRMR combines double-stranded homologous recombination [104] and molecular barcodes synthesized from DNA microarrays to generate populations of mutants that are trackable by microarray or sequencing. MAGE relies on introduction of pools of single-stranded oligonucleotides that target defined locations of the genome to introduce regulatory mutations [102] or coding modifications [105]. These and other recombineering technologies are now being developed for a variety of other organisms including gram-negative bacteria [106], lactic acid bacteria [107], *Pseudomonas syringae* [108], and *Mycobacterium tuberculosis* [109], and are likely to be very useful for engineering human-associated microbes.

3.3 Reconstituted Communities

The community of microbes that make up the human microbiome can be considered a "pseudo-organ" of its own. These microbes interact with one another and the mammalian host in potentially highly complex ways that may be difficult to decipher even with tractable genetic systems [110]. A direct approach to study these interactions is to build reconstituted communities of microbes derived from monoculture isolates in defined combinations. This de novo reconstitution approach to build synthetic communities has significant advantages over attempts to deconvolute natural communities. Reconstituted synthetic consortium presents a tractable level of complexity in terms of number of interacting microbial species that can be tracked by sequencing and predicted with quantitative models. In one such study, researchers inoculated germ-free mice with ten representative strains of the human microbiota [111]. The mice were then fed with defined diets of macronutrients consisting of proteins, fats, polysaccharides, and sugars. By tracking the abundance of the ten-member microbial consortium using high-throughput sequencing, the researchers could predict over 60 % of the variation in species abundance as a result of diet perturbations. This avenue of investigation presents a viable approach to study the human microbiome and ways to analyze synthetically engineered microbiota.

Engineered microbes have been utilized to reconstitute synthetic communities to investigate the role of metabolic exchange. One such important metabolic exchange is that of amino acids, as they are the essential constituents of proteins. Various syntrophic cross-feeding communities have been described using auxotrophic *E. coli* and yeast strains that require different amino acid supplementation for growth [112–114]. In these syntrophic systems, metabolites that are exchanged across different biosynthetic pathways promote more syntrophic growth than those that are exchanged along the same pathway, which also relates to the cost of biosynthesis of the amino acid metabolites. Amino acid exchange is likely a large player in driving metabolism of microbial communities as a substantial fraction of all microbes are missing biosynthesis of various metabolites and thus require growth on more rich and complex substrates that are found in the gut [115].

3.4 Microbial Engineering Through Synthetic Biology

New approaches are now utilizing synthetic biology to engineer human-associated microbiota to improve health and metabolism as well as to monitor and fight diseases. These efforts focus on developing genetic circuits that actuate in an engineered host cell such as *E. coli* that can sense and respond to changes to its environment and in the presence of particular pathogens. For example, to detect the human opportunistic pathogen *Pseudomonas aeruginosa*, which often causes chronic cystic fibrosis infections and colonizes the gastrointestinal tract, *E. coli* was engineered to detect the small diffusible molecule that is excreted by *P. aeruginosa* through the quorum sensing pathway [116]. An engineered synthetic circuit was placed in nonpathogenic *E. coli*, which when placed in the presence of high-density *P. aeruginosa* triggered a self-lysis program that released a narrow-spectrum bacteriocin that specifically killed the *P. aeruginosa* strain. Similar strategies have also been demonstrated to detect and respond to *Vibrio cholera* infection using engineered *E. coli* that sense autoinducer-1 (AI1) molecules from *V. cholera* quorum sensing pathway [117]. These strategies appear to yield improved survival rates against microbial pathogenesis in murine models [117]. Quorum sensing systems, which normally help microbes detect local cell density, have been further enhanced to improve robustness and performance to enable coupled short-range and long-range feedback circuits that enable microbial communication across large distances in an engineered community.

Other microbes have been successfully engineered to perform specific functions on human-associated surfaces such as the mucosal layer of the gut epithelium. Numerous diseases that occur along the intestinal tract are targets of such engineered approaches. For example, the probiotic strain *Lactococcus lactis* has been engineered to secrete recombinant human IL-10 in the gastrointestinal tract to reduce colitis [118, 119]. Other future applications of engineered probiotics include enhancing catabolism of nutrients (e.g., lactose and gluten), modulation of the immune system, and removal of pathogens by selective toxin release [116].

3.5 In Vitro Host Models

To probe and engineer the human-associated microbial community, various in vitro models have been developed, ranging from traditional batch culturing in chemostats to microfluidic systems that incorporate host cells. Single-vessel chemostats inoculated with fecal samples from healthy individuals have helped identify HGT [120] and selective bacterial colonization on different carbohydrate substrates [121, 122]. A multichamber continuous culture system mimicking spatial, nutritional, and pH properties of different GI tract regions can be used to investigate stabilization dynamics [123–125]. Similarly, the constant-depth film fermenter resembles oral biofilm [126] and has enabled studies on biofilm formation, antibiotic resistance [126], and HGT in a multispecies oral community [127, 128]. To incorporate mammalian cells in studying host–microbial interactions, organ-on-a-chip microfluidic devices have been recently used. In one version of such a system, a gut-on-a-chip device, the microfluidic channel is coated with extracellular matrix and lined by human intestinal epithelial (Caco-2) cells. This system mimics intestinal flow and peristaltic motion, recapitulates columnar epithelium polarization and intestinal villi formation, and supports the growth of commensal *Lactobacillus rhamnosus GG* [129]. These microdevices offer an opportunity to investigate host–microbiota interactions in a well-controlled manner and in physiologically relevant conditions.

Inoculating with native microbiota samples provides a method to overcome the un-cultivability of many microbes as well as to study collective activity and discover novel functions without a priori knowledge of community composition. However, starting with a predefined microbial community allows a controlled setting better suited for testing engineered systems. In one study analyzing the dynamics of a community representing the four main gut phyla in a chemostat, the authors propose that intrinsic microbial interactions, rather than host selective pressure, play a role in the observed colonization pattern, which was similar to what has been documented in the human gut [130]. Similar models have been developed for oral microbiota studies. The use of predefined oral microbial inocula has helped elucidate metabolic cooperation in batch culture [12] and community development in saliva-conditioned flow cells [131].

3.6 In Vivo Host Models

In order to move into in vivo animal models that more closely represent the physiology of the human host environment, researchers have extensively utilized murine models including germ-free, gnotobiotic, and conventionally raised mice. Gnotobiotic animals are born in aseptic conditions and reared in a sterile environment where they are exposed only to known microbial species; technically, germ-free mice are a type of gnotobiotic mice that have not been exposed to any microbes. Similar to in vitro systems, mice can be inoculated with either a natural microbiota sample or a predefined microbial community. Fecal samples, as well as oral swab and saliva samples, can then be collected from gnotobiotic mice for biochemical analysis and species quantification of gut and oral cavity microbiota. In vivo models have been used to study the transmission of antibiotic resistance in the mouse gut [132, 133] and colonization resistance in the oral cavity [134]. Furthermore, the choice of the inoculum donor offers opportunities to compare different host selection pressures and microbial community responses. Microbiota can be transplanted from conventionally raised to germ-free animals of not only the same species but also interspecies, as in human microbiota into mouse, called humanized gnotobiotic mice [134]. In one study, transplants from zebrafish gut microbiota into germ-free mice and mouse gut microbiota into germ-free zebrafish revealed that the resulting community conformed to the native host composition, demonstrating host selection [135].

Altering host diet, environment, or genetic background can also enable studies in host–microbiota interactions. One method to gain insight into the role of microbial communities in disease is to utilize mice with recapitulated pathologies. For example, IL-10$^{-/-}$, ob$^{-/-}$, apoE$^{-/-}$, and TLR2$^{-/-}$ or TLR5$^{-/-}$ mice are models for colitis, obesity, hypercholesterolemia, and metabolic syndrome, respectively [46, 136–139]. To generate antigen- or pathogen-specific phenotypes, mice can be infected with *Salmonella typhimurium* to study colitis [140] or *Citrobacter rodentium* as a model for attaching and effacing pathogens, such as enterohemorrhagic *E. coli* [141, 142]. Furthermore, murine models with chemically induced inflammation can be tools to study chronic mucosal inflammation; dextran sodium sulfate (DSS) can induce ulcerative colitis, and trinitrobenzene sulfonic acid (TNBS) can stimulate Crohn's disease [143]. To investigate oral microbiota, there are periodontal disease [144] and oral infection models [145, 146]; gnotobiotic rodents can also be fed a high-sucrose cariogenic diet to promote plaque formation.

Germ-free mice inoculated with defined microbes are informative models for analyzing microbial colonization and metabolic adaptation [147]. For example, resident bacteria and probiotic strains adapt their substrate utilization: in the presence of *Bifidobacterium longum*, *Bifidobacterium animalis*, or *Lactobacillus casei*, *Bacteroides thetaiotaomicron* diversified its carbohydrate utilization by shifting metabolism from mucosal glycans to dietary plant polysaccharides [148]. Furthermore, the effect of different diets on microbial community composition can be studied, as in gnotobiotic mice inoculated with ten sequenced gut bacterial species and fed with various levels of casein, cornstarch, sucrose, and corn oil to represent protein, polysaccharide, sugar, and fat content in the diet, respectively [111].

3.7 Computational Frameworks for Human Microbiomics

Over the past several decades, a large number of theoretical and quantitative models have been developed to describe the cell and its behavior. Constrain-based models are used to describe metabolism of individual cells using stoichiometric representation of metabolic reactions and optimization constraints [149]. Approaches such as flux balance analysis (FBA) enable the analysis of metabolism under steady-state assumptions by linear optimization solution methods. These methods are now being scaled to ecosystems of cells. Recent developments using multi-level objective optimization [150] and dynamic systems [151] enable the modeling of synthetic ecosystems of three or more members. Using metagenomic data of the gut microbiome, Greenblum et al. generated a community-level metabolic reconstruction network of the microbiota and discovered topological variations that are associated with obesity and IBD, giving rise to low diversity and differences in community composition [152]. For models that account for systems dynamics, population abundance and metabolite concentrations can be solved independently through different FBA models that are iterated at each time step. This approach called dynamic multi-species metabolic modeling (DMMM) can capture scenarios of resource competition, leading to the identification of limiting metabolites [153]. Other complementary models include elementary mode analysis (EMA) [154] that enables quantitative analysis of microbial ecosystems in a multicellular fashion.

4   Perspectives

Reframing the microbiota community as a core set of genes, not a core set of species, opens a new front to the microbiome engineering design space. In a metagenomic study of 154 individuals, no single-gut bacterial phylotype was detected at an abundant frequency amongst all the samples, a finding that is consistent with the idea that the core human gut microbiome may not be best defined by prominent species but by abundantly shared genes and functions [155]. We propose that manipulation at the gene, genome, and ultimately metagenome level offers the ability for precise multicellular engineering of desirable traits in human-associated microbiota. Besides controlled perturbations of the microbiome to advance our understanding of host–microbiota interactions, metagenome-scale tools enable novel developments in diagnostics and therapeutics.

From biosensors on the skin to reporters in the gut, there are several opportunities in monitoring the health and disease status of the human host, such as sensing nutritional deficiencies, immune imbalances, environmental toxins, or invading pathogens. Prophylactic and therapeutic avenues for human microbiome engineering include modifying community composition, tuning metabolic activity, mediating microbe–microbe relationships, and modulating host–microbe interactions. Two current microbiota-associated treatments have shown clinical efficacy: (1) fecal transplants for recurrent *Clostridium difficile* infection [156] and (2) probiotics for pouchitis, which is inflammation of the ileal pouch that is created after surgical removal of the colon in ulcerative colitis patients [157–159]. The main challenge is transmission of undesirable agents from donor feces to the recipient gut in fecal transplants and native colonization resistance that would impair infiltration and growth of new species in probiotics [160, 161]. Nevertheless, these successful approaches demonstrate the potential benefits of leveraging natural microorganisms and entire microbial communities.

In fact, coupling organismal and functional gene-level approaches would be a powerful way to engineer the native microbiota. Microbiome engineering enables multiscale system design for the synthesis of nutrients and vitamins, enhanced digestion of gluten and lactose, decreased acidity of the oral cavity, targeted elimination of multidrug-resistant pathogens, and microbial modulation of the host immune system. As vehicles for drug delivery, commensal bacteria designed to secrete heterologous genes have been explored for treating cancer [162–164], diabetes [165], HIV [166], and IBD [118]. For example, IL-10 has immunomodulatory effects in IBD but requires localized delivery at the intestinal lining to avoid the toxic side effects and low efficacy of systemic IL-10 injection. Ingestion of modified *Lactococcus lactis* that secrete recombinant IL-10 is safe and effective in animal models and has been promising in human clinical trials for IBD [119, 167].

Finally, besides addressing clinical safety and efficacy criteria for FDA regulatory approval [168], overall safety precautions are critical considerations to minimize unintentional risks in releasing genetically modified material into the natural environment. Rational design, such as creating auxotrophic microbes [119], for robust stability, non-pathogenicity, and containment of recombinant genetic systems will be essential in microbiome engineering.

Acknowledgements

H.H.W. acknowledges the generous support from the National Institutes of Health Director's Early Independence Award (grant 1DP5OD009172-01). S.J.Y. acknowledges support from the National Science Foundation Graduate Research Fellowship and the MIT Neurometrix Presidential Graduate Fellowship. G.M.C. acknowledges support from the Department of Energy Genomes to Life Center (Grant DE-FG02-02ER63445).

References

1. Huttenhower C, Gevers D, Knight R et al (2012) Structure, function and diversity of the healthy human microbiome. Nature 486:207–214

2. Ley RE, Peterson DA, Gordon JI (2006) Ecological and evolutionary forces shaping microbial diversity in the human intestine. Cell 124:837–848

3. Turnbaugh PJ, Ley RE, Hamady M et al (2007) The human microbiome project. Nature 449:804–810

4. Nicholson JK, Holmes E, Wilson ID (2005) Gut microorganisms, mammalian metabolism and personalized health care. Nat Rev Microbiol 3:431–438

5. Dethlefsen L, McFall-Ngai M, Relman DA (2007) An ecological and evolutionary perspective on human-microbe mutualism and disease. Nature 449:811–818

6. Qin J, Li R, Raes J et al (2010) A human gut microbial gene catalogue established by metagenomic sequencing. Nature 464:59–65

7. Kaeberlein T, Lewis K, Epstein SS (2002) Isolating "uncultivable" microorganisms in pure culture in a simulated natural environment. Science 296:1127–1129

8. Hayes CS, Aoki SK, Low DA (2010) Bacterial contact-dependent delivery systems. Annu Rev Genet 44:71–90

9. Bassler BL, Losick R (2006) Bacterially speaking. Cell 125:237–246

10. Walker AW, Duncan SH, Harmsen HJ et al (2008) The species composition of the human intestinal microbiota differs between particle-associated and liquid phase communities. Environ Microbiol 10:3275–3283

11. Smillie CS, Smith MB, Friedman J et al (2011) Ecology drives a global network of gene exchange connecting the human microbiome. Nature 480:241–244

12. Bradshaw DJ, Homer KA, Marsh PD et al (1994) Metabolic cooperation in oral microbial communities during growth on mucin. Microbiology 140:3407–3412

13. Falony G, Vlachou A, Verbrugghe K et al (2006) Cross-feeding between Bifido-bacterium longum BB536 and acetate-converting, butyrate-producing colon bacteria during growth on oligofructose. Appl Environ Microbiol 72:7835–7841

14. Salazar N, Gueimonde M, Hernández-Barranco AM et al (2008) Exopolysaccharides produced by intestinal Bifidobacterium strains act as fermentable substrates for human intestinal bacteria. Appl Environ Microbiol 74:4737–4745

15. Gibson GR, Cummings JH, Macfarlane GT et al (1990) Alternative pathways for hydrogen disposal during fermentation in the human colon. Gut 31:679–683

16. Dabard J, Bridonneau C, Phillipe C et al (2001) Ruminococcin A, a new lantibiotic produced by a Ruminococcus gnavus strain isolated from human feces. Appl Environ Microbiol 67:4111–4118

17. Santagati M, Scillato M, Patanè F et al (2012) Bacteriocin-producing oral streptococci and inhibition of respiratory pathogens. FEMS Immunol Med Microbiol 65:23–31

18. Gillor O, Etzion A, Riley MA (2008) The dual role of bacteriocins as anti- and probiotics. Appl Microbiol Biotechnol 81:591–606

19. Davey ME, O'toole GA (2000) Microbial biofilms: from ecology to molecular genetics. Microbiol Mol Biol Rev 64:847–867

20. Marsh PD, Moter A, Devine DA (2011) Dental plaque biofilms: communities, conflict and control. Periodontology 2000 2000(55): 16–35

21. Boles BR, Thoendel M, Singh PK (2004) Self-generated diversity produces "insurance effects" in biofilm communities. Proc Natl Acad Sci U S A 101:16630–16635

22. Stewart PS, Franklin MJ (2008) Physiological heterogeneity in biofilms. Nat Rev Microbiol 6:199–210

23. Frost LS, Leplae R, Summers AO et al (2005) Mobile genetic elements: the agents of open source evolution. Nat Rev Microbiol 3: 722–732

24. Gogarten JP, Townsend JP (2005) Horizontal gene transfer, genome innovation and evolution. Nat Rev Microbiol 3:679–687

25. Norman A, Hansen LH, Sørensen SJ (2009) Conjugative plasmids: vessels of the communal gene pool. Philos Trans R Soc Lond B Biol Sci 364:2275–2289

26. Jones BV, Marchesi JR (2007) Accessing the mobile metagenome of the human gut microbiota. Mol Biosyst 3:749–758

27. Dobrindt U, Hochhut B, Hentschel U et al (2004) Genomic islands in pathogenic and environmental microorganisms. Nat Rev Microbiol 2:414–424

28. Baquero F (2004) From pieces to patterns: evolutionary engineering in bacterial pathogens. Nat Rev Microbiol 2:510–518

29. Salyers AA (1993) Gene transfer in the mammalian intestinal tract. Curr Opin Biotechnol 4:294–298

30. Reid G, Younes JA, Van der Mei HC et al (2010) Microbiota restoration: natural and supplemented recovery of human microbial communities. Nat Rev Microbiol 9:27–38

31. Koenig JE, Spor A, Scalfone N et al (2010) Succession of microbial consortia in the developing infant gut microbiome. Proc Natl Acad Sci U S A 108(Suppl 1):4578–4585

32. Van den Abbeele P, Van de Wiele T, Verstraete W et al (2011) The host selects mucosal and luminal associations of coevolved gut microorganisms: a novel concept. FEMS Microbiol Rev 35:681–704

33. Giraud A, Arous S, De Paepe M et al (2008) Dissecting the genetic components of adaptation of Escherichia coli to the mouse gut. PLoS Genet 4:e2

34. Gill SR, Pop M, Deboy RT et al (2006) Metagenomic analysis of the human distal gut microbiome. Science 312:1355–1359

35. Bäckhed F, Ley RE, Sonnenburg JL et al (2005) Host-bacterial mutualism in the human intestine. Science 307:1915–1920

36. Guarner F, Malagelada J-R (2003) Gut flora in health and disease. Lancet 361:512–519

37. Stappenbeck TS, Hooper LV, Gordon JI (2002) Developmental regulation of intestinal angiogenesis by indigenous microbes via Paneth cells. Proc Natl Acad Sci U S A 99: 15451–15455

38. Rakoff-Nahoum S, Paglino J, Eslami-Varzaneh F et al (2004) Recognition of commensal microflora by toll-like receptors is required for intestinal homeostasis. Cell 118: 229–241

39. Hooper LV (2004) Bacterial contributions to mammalian gut development. Trends Microbiol 12:129–134

40. Pryde SE, Duncan SH, Hold GL et al (2002) The microbiology of butyrate formation in the human colon. FEMS Microbiol Lett 217:133–139

41. Round JL, Mazmanian SK (2010) Inducible Foxp3+ regulatory T-cell development by a commensal bacterium of the intestinal microbiota. Proc Natl Acad Sci U S A 107: 12204–12209

42. Wu GD, Chen J, Hoffmann C et al (2011) Linking long-term dietary patterns with gut microbial enterotypes. Science 334:105–108

43. Serino M, Luche E, Gres S et al (2012) Metabolic adaptation to a high-fat diet is associated with a change in the gut microbiota. Gut 61:543–553

44. Honda K, Littman DR (2011) The Microbiome in Infectious Disease and Inflammation. Annu Rev Immunol 30:759–795

45. Ley RE, Bäckhed F, Turnbaugh P et al (2005) Obesity alters gut microbial ecology. Proc Natl Acad Sci U S A 102:11070–11075

46. Turnbaugh PJ, Bäckhed F, Fulton L et al (2008) Diet-induced obesity is linked to marked but reversible alterations in the mouse distal gut microbiome. Cell Host Microbe 3:213–223

47. Murphy EF, Cotter PD, Healy S et al (2010) Composition and energy harvesting capacity of the gut microbiota: relationship to diet, obesity and time in mouse models. Gut 59:1635–1642

48. Cerf-Bensussan N, Gaboriau-Routhiau V (2010) The immune system and the gut microbiota: friends or foes? Nat Rev Immunol 10:735–744

49. Wen L, Ley RE, Volchkov PY et al (2008) Innate immunity and intestinal microbiota in the development of Type 1 diabetes. Nature 455:1109–1113

50. Lee YK, Menezes JS, Umesaki Y et al (2010) Proinflammatory T-cell responses to gut microbiota promote experimental autoimmune encephalomyelitis. Proc Natl Acad Sci U S A 108(Suppl 1):4615–4622

51. Abraham C, Cho JH (2009) Inflammatory bowel disease. N Engl J Med 361:2066–2078

52. Nell S, Suerbaum S, Josenhans C (2010) The impact of the microbiota on the pathogenesis of IBD: lessons from mouse infection models. Nat Rev Microbiol 8:564–577

53. Sokol H, Seksik P, Furet JP et al (2009) Low counts of faecalibacterium prausnitzii in colitis microbiota. Inflamm Bowel Dis 15:1183–1189

54. Manichanh C, Rigottier-Gois L, Bonnaud E et al (2006) Reduced diversity of faecal microbiota in Crohn's disease revealed by a metagenomic approach. Gut 55:205–211

55. He T, Venema K, Priebe MG et al (2008) The role of colonic metabolism in lactose intolerance. Eur J Clin Invest 38:541–547

56. He T, Priebe MG, Harmsen HJM et al (2006) Colonic fermentation may play a role in lactose intolerance in humans. J Nutr 136:58

57. Tehrani AB, Nezami BG, Gewirtz A et al (2012) Obesity and its associated disease: a role for microbiota? Neurogastroenterol Motil 24:305–311

58. Everard A, Lazarevic V, Derrien M et al (2011) Responses of gut microbiota and glucose and lipid metabolism to prebiotics in genetic obese and diet-induced leptin-resistant mice. Diabetes 60:2775–2786

59. Giongo A, Gano KA, Crabb DB et al (2010) Toward defining the autoimmune microbiome for type 1 diabetes. ISME J 5:82–91

60. Wu H-J, Ivanov II, Darce J et al (2010) Gut-residing segmented filamentous bacteria drive autoimmune arthritis via T helper 17 cells. Immunity 32:815–827

61. Lam V, Su J, Koprowski S et al (2012) Intestinal microbiota determine severity of myocardial infarction in rats. FASEB J 26(4): 1727–1735

62. Wardwell LH, Huttenhower C, Garrett WS (2011) Current concepts of the intestinal microbiota and the pathogenesis of infection. Curr Infect Dis Rep 13:28–34

63. Gori A, Tincati C, Rizzardini G et al (2008) Early impairment of gut function and gut flora supporting a role for alteration of gastro-intestinal mucosa in human immunodeficiency virus pathogenesis. J Clin Microbiol 46:757–758

64. Stecher B, Hardt W-D (2008) The role of microbiota in infectious disease. Trends Microbiol 16:107–114

65. Walk ST, Young VB (2008) Emerging insights into antibiotic-associated diarrhea and clostridium difficile infection through the lens of microbial ecology. Interdiscip Perspect Infect Dis 2008:125081

66. Vrieze A, Holleman F, Zoetendal EG et al (2010) The environment within: how gut microbiota may influence metabolism and body composition. Diabetologia 53:606–613

67. Hou JK, Abraham B, El-Serag H (2011) Dietary intake and risk of developing inflammatory bowel disease: a systematic review of the literature. Am J Gastroenterol 106: 563–573

68. Fava F, Lovegrove JA, Gitau R et al (2006) The gut microbiota and lipid metabolism: implications for human health and coronary heart disease. Curr Med Chem 13:3005–3021

69. Wang Z, Klipfell E, Bennett BJ et al (2011) Gut flora metabolism of phosphatidylcholine promotes cardiovascular disease. Nature 472:57–63

70. Dobkin JF, Saha JR, Butler VP et al (1983) Digoxin-inactivating bacteria: identification in human gut flora. Science 220:325–327

71. Clayton TA, Baker D, Lindon JC et al (2009) Pharmacometabonomic identification of a significant host-microbiome metabolic interaction affecting human drug metabolism. Proc Natl Acad Sci U S A 106:14728–14733

72. Wallace BD, Wang H, Lane KT et al (2010) Alleviating cancer drug toxicity by inhibiting a bacterial enzyme. Science 330:831–835

73. Marsh PD (1994) Microbial ecology of dental plaque and its significance in health and disease. Adv Dent Res 8:263–271

74. Azarpazhooh A, Leake JL (2006) Systematic review of the association between respiratory diseases and oral health. J Periodontol 77:1465–1482

75. Ford PJ, Gemmell E, Timms P et al (2007) Anti-P. gingivalis response correlates with atherosclerosis. J Dent Res 86:35–40

76. Li L, Messas E, Batista EL et al (2002) Porphyromonas gingivalis infection accelerates the progression of atherosclerosis in a heterozygous apolipoprotein e-deficient murine model. Circulation 105:861–867

77. Koren O, Spor A, Felin J et al (2010) Human oral, gut, and plaque microbiota in patients with atherosclerosis. Proc Natl Acad Sci U S A 108(Suppl 1):4592–4598

78. Haug MC, Tanner SA, Lacroix C et al (2011) Monitoring horizontal antibiotic resistance gene transfer in a colonic fermentation model. FEMS Microbiol Ecol 78:210–219

79. Nelson KE, Weinstock GM, Highlander SK et al (2010) A catalog of reference genomes from the human microbiome. Science 328: 994–999

80. Methe BA, Nelson KE, Pop M et al (2012) A framework for human microbiome research. Nature 486:215–221

81. Park J, Kerner A, Burns MA et al (2011) Microdroplet-enabled highly parallel co-cultivation of microbial communities. PLoS One 6:e17019

82. Bollmann A, Lewis K, Epstein SS (2007) Incubation of environmental samples in a diffusion chamber increases the diversity of recovered isolates. Appl Environ Microbiol 73:6386–6390

83. Thomas CM, Nielsen KM (2005) Mechanisms of, and barriers to, horizontal gene transfer between bacteria. Nat Rev Microbiol 3: 711–721

84. Lorenz MG, Wackernagel W (1994) Bacterial gene transfer by natural genetic transformation in the environment. Microbiol Rev 58:563–602

85. Wirth R, Friesenegger A, Fiedler S (1989) Transformation of various species of gram-negative bacteria belonging to 11 different genera by electroporation. Mol Gen Genet 216:175–177

86. Sanford JC, Smith FD, Russell JA (1993) Optimizing the biolistic process for different biological applications. Methods Enzymol 217:483–509

87. Wyber JA, Andrews J, D'Emanuele A (1997) The use of sonication for the efficient delivery of plasmid DNA into cells. Pharm Res 14: 750–756

88. Swords WE (2003) Chemical transformation of E. coli. Methods Mol Biol 235:49–53

89. Thomson AM, Flint HJ (1989) Electroporation induced transformation of Bacteroides ruminicola and Bacteroides uniformis by plasmid DNA. FEMS Microbiol Lett 52:101–104

90. Calvin NM, Hanawalt PC (1988) High-efficiency transformation of bacterial cells by electroporation. J Bacteriol 170:2796–2801

91. Goodman AL, McNulty NP, Zhao Y et al (2009) Identifying genetic determinants needed to establish a human gut symbiont in its habitat. Cell Host Microbe 6:279–289

92. Kleckner N (1981) Transposable elements in prokaryotes. Annu Rev Genet 15:341–404

93. Goodman AL, Wu M, Gordon JI (2011) Identifying microbial fitness determinants by insertion sequencing using genome-wide transposon mutant libraries. Nat Protoc 6:1969–1980

94. van Opijnen T, Bodi KL, Camilli A (2009) Tn-seq: high-throughput parallel sequencing for fitness and genetic interaction studies in microorganisms. Nat Methods 6:767–772

95. Gawronski JD, Wong SM, Giannoukos G et al (2009) Tracking insertion mutants within libraries by deep sequencing and a genome-wide screen for Haemophilus genes required in the lung. Proc Natl Acad Sci U S A 106:16422–16427

96. Langridge GC, Phan MD, Turner DJ et al (2009) Simultaneous assay of every Salmonella Typhi gene using one million transposon mutants. Genome Res 19:2308–2316

97. Sommer MO, Dantas G, Church GM (2009) Functional characterization of the antibiotic resistance reservoir in the human microflora. Science 325:1128–1131

98. Warner JR, Reeder PJ, Karimpour-Fard A et al (2010) Rapid profiling of a microbial genome using mixtures of barcoded oligonucleotides. Nat Biotechnol 28:856–862

99. Sandoval NR, Kim JY, Glebes TY et al (2012) Strategy for directing combinatorial genome engineering in Escherichia coli. Proc Natl Acad Sci U S A 109:10540–10545

100. Wang HH, Isaacs FJ, Carr PA et al (2009) Programming cells by multiplex genome engineering and accelerated evolution. Nature 460:894–898

101. Wang HH, Church GM (2011) Multiplexed genome engineering and genotyping methods applications for synthetic biology and metabolic engineering. Methods Enzymol 498:409–426

102. Wang HH, Kim H, Cong L et al (2012) Genome-scale promoter engineering by coselection MAGE. Nat Methods 9:591–593

103. Carr PA, Wang HH, Sterling B et al (2012) Enhanced multiplex genome engineering through co-operative oligonucleotide co-selection. Nucleic Acids Res 40:e132

104. Sharan SK, Thomason LC, Kuznetsov SG et al (2009) Recombineering: a homologous recombination-based method of genetic engineering. Nat Protoc 4:206–223

105. Isaacs FJ, Carr PA, Wang HH et al (2011) Precise manipulation of chromosomes in vivo enables genome-wide codon replacement. Science 333:348–353

106. Swingle B, Markel E, Costantino N et al (2010) Oligonucleotide recombination in Gram-negative bacteria. Mol Microbiol 75:138–148

107. van Pijkeren J-P, Britton RA (2012) High efficiency recombineering in lactic acid bacteria. Nucleic Acids Res 40:e76

108. Swingle B, Bao Z, Markel E et al (2010) Recombineering using RecTE from Pseudomonas syringae. Appl Environ Microbiol 76:4960–4968

109. van Kessel JC, Hatfull GF (2007) Recombineering in Mycobacterium tuberculosis. Nat Methods 4:147–152

110. Sonnenburg JL, Angenent LT, Gordon JI (2004) Getting a grip on things: how do communities of bacterial symbionts become established in our intestine? Nat Immunol 5:569–573

111. Faith JJ, McNulty NP, Rey FE et al (2011) Predicting a human gut microbiota's response to diet in gnotobiotic mice. Science 333:101–104

112. Hosoda K, Suzuki S, Yamauchi Y et al (2011) Cooperative adaptation to establishment of a synthetic bacterial mutualism. PLoS One 6:e17105

113. Shou W, Ram S, Vilar JM (2007) Synthetic cooperation in engineered yeast populations. Proc Natl Acad Sci U S A 104:1877–1882

114. Wintermute EH, Silver PA (2010) Emergent cooperation in microbial metabolism. Mol Syst Biol 6:407

115. Mee JM, Wang HH (2012) Engineering ecosystems and synthetic ecologies. Mol Biosyst 8:2470–2483

116. Saeidi N, Wong CK, Lo TM et al (2011) Engineering microbes to sense and eradicate Pseudomonas aeruginosa, a human pathogen. Mol Syst Biol 7:521

117. Duan F, March JC (2010) Engineered bacterial communication prevents Vibrio cholerae

virulence in an infant mouse model. Proc Natl Acad Sci U S A 107:11260–11264

118. Steidler L (2000) Treatment of murine colitis by lactococcus lactis secreting interleukin-10. Science 289:1352–1355

119. Steidler L, Rottiers P, Coulie B (2009) Actobiotics as a novel method for cytokine delivery. Ann N Y Acad Sci 1182:135–145

120. Duncan SH, Scott KP, Ramsay AG et al (2003) Effects of alternative dietary substrates on competition between human colonic bacteria in an anaerobic fermentor system. Appl Environ Microbiol 69:1136–1142

121. Leitch ECM, Walker AW, Duncan SH et al (2007) Selective colonization of insoluble substrates by human faecal bacteria. Environ Microbiol 9:667–679

122. Macfarlane GT, Hay S, Gibson GR (1989) Influence of mucin on glycosidase, protease and arylamidase activities of human gut bacteria grown in a 3-stage continuous culture system. J Appl Bacteriol 66:407–417

123. Molly K, Woestyne M, Verstraete W (1993) Development of a 5-step multi-chamber reactor as a simulation of the human intestinal microbial ecosystem. Appl Microbiol Biotechnol 39:254–258

124. Possemiers S, Verthé K, Uyttendaele S et al (2004) PCR-DGGE-based quantification of stability of the microbial community in a simulator of the human intestinal microbial ecosystem. FEMS Microbiol Ecol 49:495–507

125. Pratten J (2007) Growing oral biofilms in a constant depth film fermentor (CDFF). *Curr Protoc Microbiol* Chapter 1, Unit 1B.5

126. Ready D (2002) Composition and antibiotic resistance profile of microcosm dental plaques before and after exposure to tetracycline. J Antimicrob Chemother 49:769–775

127. Roberts AP, Pratten J, Wilson M et al (1999) Transfer of a conjugative transposon, Tn5397 in a model oral biofilm. FEMS Microbiol Lett 177:63–66

128. Roberts AP, Cheah G, Ready D et al (2001) Transfer of Tn916-like elements in microcosm dental plaques. Antimicrob Agents Chemother 45:2943–2946

129. Kim HJ, Huh D, Hamilton G et al (2012) Human Gut-on-a-Chip inhabited by microbial flora that experiences intestinal peristalsis-like motions and flow. Lab Chip 12:2165–2174

130. Trosvik P, Rudi K, Strætkvern KO et al (2010) Web of ecological interactions in an experimental gut microbiota. Environ Microbiol 12:2677–2687

131. Foster JS, Kolenbrander PE (2004) Development of a multispecies oral bacterial community in a saliva-conditioned flow cell. Appl Environ Microbiol 70:4340

132. Doucet-Populaire F, Trieu-Cuot P, Dosbaa I et al (1991) Inducible transfer of conjugative transposon Tn1545 from Enterococcus faecalis to Listeria monocytogenes in the digestive tracts of gnotobiotic mice. Antimicrob Agents Chemother 35:185–187

133. Launay A, Ballard SA, Johnson PDR et al (2006) Transfer of vancomycin resistance transposon Tn1549 from clostridium symbiosum to Enterococcus spp. in the gut of gnotobiotic mice. Antimicrob Agents Chemother 50:1054

134. Turnbaugh PJ, Ridaura VK, Faith JJ et al (2009) The effect of diet on the human gut microbiome: a metagenomic analysis in humanized gnotobiotic mice. Sci Transl Med 1:6ra14

135. Rawls JF, Mahowald MA, Ley RE et al (2006) Reciprocal gut microbiota transplants from zebrafish and mice to germ-free recipients reveal host habitat selection. Cell 127:423–433

136. Sellon RK, Tonkonogy S, Schultz M et al (1998) Resident enteric bacteria are necessary for development of spontaneous colitis and immune system activation in interleukin-10-deficient mice. Infect Immun 66:5224–5231

137. Lalla E, Lamster IB, Hofmann MA et al (2003) Oral infection with a periodontal pathogen accelerates early atherosclerosis in apolipoprotein E-null mice. Arterioscler Thromb Vasc Biol 23:1405–1411

138. Caricilli AM, Picardi PK, de Abreu LL et al (2011) Gut microbiota is a key modulator of insulin resistance in TLR 2 knockout mice. PLoS Biol 9:e1001212

139. Vijay-Kumar M, Aitken JD, Carvalho FA et al (2010) Metabolic syndrome and altered gut microbiota in mice lacking Toll-like receptor 5. Science 328:228–231

140. Hapfelmeier S, Hardt W-D (2005) A mouse model for S. typhimurium-induced enterocolitis. Trends Microbiol 13:497–503

141. Deng W, Vallance BA, Li Y et al (2003) Citrobacter rodentium translocated intimin receptor (Tir) is an essential virulence factor needed for actin condensation, intestinal colonization and colonic hyperplasia in mice. Mol Microbiol 48:95–115

142. Newman JV, Zabel BA, Jha SS et al (1999) Citrobacter rodentium espB is necessary for signal transduction and for infection of laboratory mice. Infect Immun 67:6019–6025

143. Alex P, Zachos NC, Nguyen T et al (2009) Distinct cytokine patterns identified from

multiplex profiles of murine DSS and TNBS-induced colitis. Inflamm Bowel Dis 15:341–352

144. Oz HS, Puleo DA (2011) Animal models for periodontal disease. J Biomed Biotechnol 2011:1–8

145. Naglik JR, Fidel PL, Odds FC (2008) Animal models of mucosal Candida infection. FEMS Microbiol Lett 283:129–139

146. Mcbride BC, van der Hoeven JS (1981) Role of interbacterial adherence in colonization of the oral cavities of gnotobiotic rats infected with Streptococcus mutans and Veillonella alcalescens. Infect Immun 33:467–472

147. Ma M, Rey FE, Seedorf H et al (2009) Characterizing a model human gut microbiota composed of members of its two dominant bacterial phyla. Proc Natl Acad Sci U S A 106:5859–5864

148. Sonnenburg JL, Chen CTL, Gordon JI (2006) Genomic and metabolic studies of the impact of probiotics on a model gut symbiont and host. PLoS Biol 4:e413

149. Lewis NE, Nagarajan H, Palsson BO (2012) Constraining the metabolic genotype-phenotype relationship using a phylogeny of in silico methods. Nat Rev Microbiol 10:291–305

150. Zomorrodi AR, Maranas CD (2012) OptCom: a multi-level optimization framework for the metabolic modeling and analysis of microbial communities. PLoS Comput Biol 8:e1002363

151. Mahadevan R, Edwards JS, Doyle FJ 3rd (2002) Dynamic flux balance analysis of diauxic growth in Escherichia coli. Biophys J 83:1331–1340

152. Greenblum S, Turnbaugh PJ, Borenstein E (2012) Metagenomic systems biology of the human gut microbiome reveals topological shifts associated with obesity and inflammatory bowel disease. Proc Natl Acad Sci U S A 109:594–599

153. Zhuang K, Izallalen M, Mouser P et al (2011) Genome-scale dynamic modeling of the competition between Rhodoferax and Geobacter in anoxic subsurface environments. ISME J 5:305–316

154. Taffs R, Aston JE, Brileya K et al (2009) In silico approaches to study mass and energy flows in microbial consortia: a syntrophic case study. BMC Syst Biol 3:114

155. Turnbaugh PJ, Hamady M, Yatsunenko T et al (2009) A core gut microbiome in obese and lean twins. Nature 457:480–484

156. Rohlke F, Surawicz CM, Stollman N (2010) Fecal flora reconstitution for recurrent Clostridium difficile infection: results and methodology. J Clin Gastroenterol 44:567–570

157. Miele E, Pascarella F, Giannetti E et al (2009) Effect of a probiotic preparation (VSL#3) on induction and maintenance of remission in children with ulcerative colitis. Am J Gastroenterol 104:437–443

158. Gionchetti P, Rizzello F, Helwig U et al (2003) Prophylaxis of pouchitis onset with probiotic therapy: a double-blind, placebo-controlled trial. Gastroenterology 124:1202–1209

159. Mimura T, Rizzello F, Helwig U et al (2004) Once daily high dose probiotic therapy (VSL#3) for maintaining remission in recurrent or refractory pouchitis. Gut 53:108–114

160. Culligan EP, Hill C, Sleator RD (2009) Probiotics and gastrointestinal disease: successes, problems and future prospects. Gut Pathogens 1:19

161. Sartor RB (2004) Therapeutic manipulation of the enteric microflora in inflammatory bowel diseases: antibiotics, probiotics, and prebiotics. Gastroenterology 126:1620–1633

162. Cronin M, Morrissey D, Rajendran S et al (2010) Orally administered bifidobacteria as vehicles for delivery of agents to systemic tumors. Mol Ther 18:1397–1407

163. Fu G-F, Li X, Hou Y-Y et al (2005) Bifidobacterium longum as an oral delivery system of endostatin for gene therapy on solid liver cancer. Cancer Gene Ther 12:133–140

164. Li X, Fu G-F, Fan Y-R et al (2003) Bifidobacterium adolescentis as a delivery system of endostatin for cancer gene therapy: selective inhibitor of angiogenesis and hypoxic tumor growth. Cancer Gene Ther 10:105–111

165. Duan F, Curtis KL, March JC (2008) Secretion of insulinotropic proteins by commensal bacteria: rewiring the gut to treat diabetes. Appl Environ Microbiol 74:7437–7438

166. Rao S, Hu S, McHugh L et al (2005) Toward a live microbial microbicide for HIV: commensal bacteria secreting an HIV fusion inhibitor peptide. Proc Natl Acad Sci U S A 102:11993–11998

167. Braat H, Rottiers P, Hommes DW et al (2006) A phase I trial with transgenic bacteria expressing interleukin-10 in Crohn's disease. Clin Gastroenterol Hepatol 4:754–759

168. Degnan FH (2008) The US Food and Drug Administration and probiotics: regulatory categorization. Clin Infect Dis 46(Suppl 2): S133–S136, discussion S144–S151

169. Hong P-Y, Lee BW, Aw M et al (2010) Comparative analysis of fecal microbiota in infants with and without eczema. PLoS One 5:e9964

170. Saulnier DM, Riehle K, Mistretta T-A et al (2011) Gastrointestinal microbiome signatures of pediatric patients with irritable bowel syndrome. Gastroenterology 141:1782–1791

171. Claesson MJ, Cusack S, O'Sullivan O et al (2011) Composition, variability, and temporal stability of the intestinal microbiota of the elderly. Proc Natl Acad Sci U S A 108(Suppl):4586–4591

172. Yatsunenko T, Rey FE, Manary MJ et al (2012) Human gut microbiome viewed across age and geography. Nature 486:222–227

173. Spor A, Koren O, Ley R (2011) Unravelling the effects of the environment and host genotype on the gut microbiome. Nat Rev Microbiol 9:279–290

174. De Filippo C, Cavalieri D, Di Paola M et al (2010) Impact of diet in shaping gut microbiota revealed by a comparative study in children from Europe and rural Africa. Proc Natl Acad Sci U S A 107:14691–14696

175. Peterson DA, Frank DN, Pace NR et al (2008) Metagenomic approaches for defining the pathogenesis of inflammatory bowel diseases. Cell Host Microbe 3:417–427

176. Larsen N, Vogensen FK, van den Berg FWJ et al (2010) Gut microbiota in human adults with type 2 diabetes differs from non-diabetic adults. PLoS One 5:e9085

177. Kong HH, Oh J, Deming C et al (2012) Temporal shifts in the skin microbiome associated with atopic dermatitis disease flares and treatment. Genome Res 22(5):850–859

178. Gao Z, C-h T, Pei Z et al (2007) Molecular analysis of human forearm superficial skin bacterial biota. Proc Natl Acad Sci U S A 104:2927–2932

179. Keijser BJF, Zaura E, Huse SM et al (2008) Pyrosequencing analysis of the Oral Microflora of healthy adults. J Dent Res 87:1016–1020

180. Yang F, Zeng X, Ning K et al (2012) Saliva microbiomes distinguish caries-active from healthy human populations. ISME J 6:1–10

181. Phillips-Jones MK (1995) Introduction of recombinant DNA into Clostridium spp. Methods Mol Biol 47:227–235

182. Bouillaut L, McBride SM, Sorg JA (2011) Genetic manipulation of Clostridium difficile. *Curr Protoc Microbiol* Chapter 9, Unit 9A.2

183. Jennert KC, Tardif C, Young DI et al (2000) Gene transfer to Clostridium cellulolyticum ATCC 35319. Microbiology 146(Pt 12):3071–3080

184. Young DI, Evans VJ, Jefferies JR et al (1999) Genetic methods in clostridia. Method Microbiol 29:191–207

185. Cocconcelli PS, Ferrari E, Rossi F et al (1992) Plasmid transformation of Ruminococcus albus by means of high-voltage electroporation. FEMS Microbiol Lett 73:203–207

186. Damelin LH, Mavri-Damelin D, Klaenhammer TR et al (2010) Plasmid transduction using bacteriophage Phi(adh) for expression of CC chemokines by Lactobacillus gasseri ADH. Appl Environ Microbiol 76:3878–3885

187. Lizier M, Sarra PG, Cauda R et al (2010) Comparison of expression vectors in Lactobacillus reuteri strains. FEMS Microbiol Lett 308:8–15

188. Ljungh A, Wadström T (eds) (2009) Lactobacillus molecular biology: from genomics to probiotics. Caister Academic Press, Norfolk, UK

189. Sørvig E, Mathiesen G, Naterstad K et al (2005) High-level, inducible gene expression in Lactobacillus sakei and Lactobacillus plantarum using versatile expression vectors. Microbiology 151:2439–2449

190. Thompson K, Collins MA (1996) Improvement in electroporation efficiency for Lactobacillus plantarum by the inclusion of high concentrations of glycine in the growth medium. J Microbiol Methods 26:73–79

191. Shepard BD, Gilmore MS (1995) Electroporation and efficient transformation of Enterococcus faecalis grown in high concentrations of glycine. Methods Mol Biol 47:217–226

192. Holo H, Nes IF (1995) Transformation of Lactococcus by electroporation. Methods Mol Biol 47:195–199

193. Biswas I, Jha JK, Fromm N (2008) Shuttle expression plasmids for genetic studies in Streptococcus mutans. Microbiology 154: 2275–2282

194. McLaughlin RE, Ferretti JJ (1995) Electrotransformation of Streptococci. Methods Mol Biol 47:185–193

195. Lee JC (1995) Electrotransformation of Staphylococci. Methods Mol Biol 47: 209–216

196. Alexander JE, Andrew PW, Jones D et al (1990) Development of an optimized system for electroporation of Listeria species. Lett Appl Microbiol 10:179–181

197. Kuramitsu HK, Chi B, Ikegami A (2005) Genetic manipulation of Treponema denticola. *Curr Protoc Microbiol* Chapter 12, Unit 12B.12

198. Hyde JA, Weening EH, Skare JT (2011) Genetic transformation of borrelia burgdorferi. Curr Protoc Microbiol, Chapter 12, 1–17

199. Rosa P, Stevenson B, Tilly K (1999) Genetic methods in Borrelia and other spirochaetes. Method Microbiol 29:209–227

200. Mayo B, van Sinderen D (2010) Bifidobacteria: genomics and molecular aspects. Caister Academic Press, Norfolk, UK

201. Yeung MK, Kozelsky CS (1994) Transformation of Actinomyces spp. by a gram-negative broad-host-range plasmid. J Bacteriol 176:4173–4176

202. Miles R, Nicholas R (eds) (1998) Mycoplasma protocols, vol 104, Methods Mol Biol. Humana Press, Totowa, NJ

203. Sassetti CM, Boyd DH, Rubin EJ (2001) Comprehensive identification of conditionally essential genes in mycobacteria. Proc Natl Acad Sci U S A 98:12712–12717

204. Luijk NV, Stierli MP, Schwenninger SM (2002) Genetics and molecular biology of propionibacteria. Lait 82:45–57

205. Binet R, Maurelli AT (2009) Transformation and isolation of allelic exchange mutants of Chlamydia psittaci using recombinant DNA introduced by electroporation. Proc Natl Acad Sci U S A 106:292–297

206. Bélanger M, Rodrigues P, Progulske-Fox A (2007) Genetic manipulation of Porphyromonas gingivalis. *Curr Protoc Microbiol* Chapter 13, Unit 13C.12

207. Flint HJ, Martin JC, Thomson AM (2000) Prevotella bryantii, P. ruminicola and bacteroides strains. In: Eynard N, Teissié J (eds) Electrotransformation of bacteria. Springer, Heidelberg, pp 140–149

208. Nikolich MP, Salyers AA, Shoemaker NB (1994) Method and materials for introducing dna into prevotella ruminicola. US Patent 5322784, Jun 21, 1994

209. Bacic MK, Smith CJ (2008) Laboratory maintenance and cultivation of bacteroides species. *Curr Protoc Microbiol* Chapter 13, Unit 13C 11

210. Salyers AA, Shoemaker N, Cooper A et al (1999) Genetic methods for bacteroides species. Method Microbiol 29:229–249

211. Smith CJ (1995) Genetic transformation of Bacteroides spp. using electroporation. Methods Mol Biol 47:161–169

212. Kinder Haake S, Yoder S, Gerardo SH (2006) Efficient gene transfer and targeted mutagenesis in Fusobacterium nucleatum. Plasmid 55:27–38

213. Segal ED (1995) Electroporation of Helicobacter pylori. Methods Mol Biol 47:179–184

214. Taylor DE (1992) Genetics of campylobacter and helicobacter. Annu Rev Microbiol 46:35–64

215. Rachek LI, Hines A, Tucker AM et al (2000) Transformation of Rickettsia prowazekii to erythromycin resistance encoded by the Escherichia coli ereB gene. J Bacteriol 182:3289–3291

216. McQuiston JR, Schurig GG, Sriranganathan N et al (1995) Transformation of Brucella species with suicide and broad host-range plasmids. Methods Mol Biol 47:143–148

217. Scarlato V, Ricci S, Rappuoli R et al (1996) Genetic manipulation of bordetella. In: Adolph KW (ed) Microbial genome methods. CRC Press, Boca Raton, FL, pp 247–262

218. Bogdan JA, Minetti CA, Blake MS (2002) A one-step method for genetic transformation of non-piliated Neisseria meningitidis. J Microbiol Methods 49:97–101

219. Genco CA, Knapp JS, Clark VL (1984) Conjugation of plasmids of neisseria gonorrhoeae to other neisseria species: potential reservoirs for the β-lactamase plasmid. J Infect Dis 150:397–401

220. O'Dwyer CA, Langford PR, Kroll JS (2005) A novel neisserial shuttle plasmid: a useful new tool for meningococcal research. FEMS Microbiol Lett 251:143–147

221. Dennis JJ, Sokol PA (1995) Electrotransformation of Pseudomonas. Methods Mol Biol 47:125–133

Chapter 2

Constructing Synthetic Microbial Communities to Explore the Ecology and Evolution of Symbiosis

Adam James Waite and Wenying Shou

Abstract

Synthetically engineered microbial communities based on model organisms provide a simplified model of their naturally occurring counterparts while still retaining essential features of living organisms. The degree of control afforded by this approach has been critical in understanding how similar types of natural communities might have persisted and evolved. Here, we first discuss important considerations when designing a synthetically engineered system. Then, we describe the steps required to create a two-partner cooperative system based on the yeast *Saccharomyces cerevisiae*.

Key words Evolution, Ecology, Mutualism, Cooperation, Synthetic biology, *S. cerevisiae*

1 Introduction

From mediating biogeochemical cycles [1] to influencing human health [2] and disease [3], microbial communities impact all aspects of life on earth. However, the complexity of microbial communities and the difficulty in isolating and culturing microbes [4] pose serious challenges for decoding cell–cell and cell–environment interactions. Moreover, the evolutionary histories of microbial communities are difficult to retrace. Alternatively, communities of model organisms engineered to engage in defined interactions can be deployed to address fundamental questions in ecology and evolution, such as how species coexist and coevolve [5]. In this chapter, we discuss several considerations when designing a synthetic community, using the construction of a two-partner cooperative yeast system as an example. Then, we describe the methodology in detail.

The initial consideration is the choice of organisms. Genetic tractability and short generation times facilitate strain construction and experimentation, as well as the discovery and interpretation of mutations during experimental evolution. Thus, well-studied model organisms with reference genome sequences such as *Escherichia coli*

Lianhong Sun and Wenying Shou (eds.), *Engineering and Analyzing Multicellular Systems: Methods and Protocols*, Methods in Molecular Biology, vol. 1151, DOI 10.1007/978-1-4939-0554-6_2, © Springer Science+Business Media New York 2014

or *Saccharomyces cerevisiae* are ideal, although other species have also been used [5]. While each model organism has its advantages and disadvantages, we will use *S. cerevisiae* to highlight principles that are applicable to any synthetically engineered community.

Sexual recombination is both a help and a hindrance to synthetically engineered communities. On the one hand, sexual recombination can radically simplify strain construction by allowing genetic shuffling. For example, when a strain with genotype *AB* is crossed with a strain of genotype *ab*, recombinant strains *Ab* and *aB* can be generated. On the other hand, distinct populations in an engineered community should remain genetically insulated from one another and therefore should not be allowed to mate. An advantage of using *S. cerevisiae* is its ability to undergo both sexual and asexual reproduction. Haploid yeast is particularly suitable for evolution experiments since phenotypes arising from recessive mutations are immediately apparent. Haploid yeast can reproduce asexually as either of the two mating types, "a" or "α." Two cells of opposite mating type can mate to produce an a/α diploid, which can reproduce either asexually as diploids or sexually to form haploids. The final strains for an engineered community should always be the same mating type to prevent sexual recombination. However, haploid yeast switch mating types spontaneously at very low frequency even when the gene required for mating-type switching (*HO*) is defective (which is the case for all commonly used laboratory strains). Thus, we have used *MAT*a cells in which the *STE3* gene encoding the receptor for *MAT*a mating pheromone [6] is deleted. Thus, if a *MAT*a *ste3* cell switches mating type to *MAT*α *ste3*, it will fail to initiate the mating process.

Ideally, all strains should be derived from an isogenic background so that the only mutations are the ones defined by the researcher. We have used the strain S288C, which is one of the common laboratory strains, and the wild vineyard isolate RM11-1a (hereafter referred to as "RM11"). While S288C has its genome fully annotated and easily accessible [7], it has a tendency to produce mitochondrially deficient "petite" cells [8], which are prone to nuclear genome instability [9] and could potentially interfere with evolution experiments. RM11 (genome sequence available at http://www.broadinstitute.org/annotation/genome/saccharomyces_cerevisiae/) grows faster than S288C and produces very few petites, but haploid daughter cells do not separate well from their mothers unless the RM-11 *AMN1* allele is replaced by the *AMN1* allele from S288C [10]. Genetic manipulation in RM-11 is more difficult due to its lower transformation efficiency compared to S288C.

A major advantage of using model organisms is that they are genetically modifiable. Foreign DNA can be transformed into yeast as autonomously replicating and segregating circular plasmids or as linear DNA if genomic integration is desired. Integration is more stable than using a plasmid. Different populations can be marked with, for example, different antibiotic resistances or fluorescent proteins. Currently, at least six dominant antibiotic resistance genes are in wide use with *S. cerevisiae*: *kan*, which confers resistance to geneticin a.k.a. G418; *hph*, which confers resistance to hygromycin B; *nat*, which confers resistance to nourseothricin (sold as clonNAT by Werner BioAgents); *pat*, which confers resistance to phosphinothricin; *ble*, which confers resistance to phleomycin; and *AUR1-C*, which confers resistance to Aureobasidin A (AbA). All of these genes are available on plasmids [11–13]. Our lab has used G418, hygromycin B, clonNAT, and AbA resistance markers. Using antibiotic resistance to select a subpopulation from a co-culture is especially useful if the subpopulation is very rare, as tens of millions of cells can be assayed on a single plate. However, plating on media supplemented with different drugs to determine subpopulation abundance is time-delayed since it takes at least 1 day for colonies to grow up. It is also of low throughput due to the small number (hundreds) of individual colonies that can be counted on a plate. In contrast, fluorescently tagged strains can be distinguished using flow cytometry, which allows tens of thousands of cells to be counted in less than a minute. While fluorescence-activated cell sorting (FACS) can also be used to isolate subpopulations, it requires an expensive instrument and is less efficient when isolating very rare subpopulations. A large number of fluorescent proteins are available [14], although not all of them are bright and/or resolvable from one another using standard filter sets. In addition, the correct folding of fluorescent proteins requires oxygen [15] and is therefore incompatible with strict anaerobes. We have C-terminally tagged (Fig. 1) the highly abundant proteins *FBA1* or *MET6* with different fluorescent proteins in yeast. Using the appropriate combination of lasers and filter sets (Table 1), we can resolve mixtures of cells expressing five different fluorescent proteins: CFP, GFP, YFP, mOrange, and mCherry as well as the far-red nucleic acid dye TO-PRO-3 (Invitrogen) which can be used to stain dead cells. Plasmids containing different combinations of fluorescent proteins and selectable markers (i.e., genes for nutrient biosynthesis or antibiotic resistance) [16] are readily available from EUROSCARF (http://web.uni-frankfurt.de/fb15/mikro/euroscarf/).

Since deleting or inserting genes will usually be accomplished by transformations (Fig. 1), which require selectable markers, it is convenient to be able to reuse these markers. Plasmids containing *kanMX* [17] and *ble* [18] flanked by loxP sites are available for this purpose (for a comprehensive list of plasmids with removable markers, *see* ref. 19). In the presence of Cre recombinase, the two loxP sites recombine, removing the intervening marker and allowing for its reuse in another round of manipulation. Our lab has

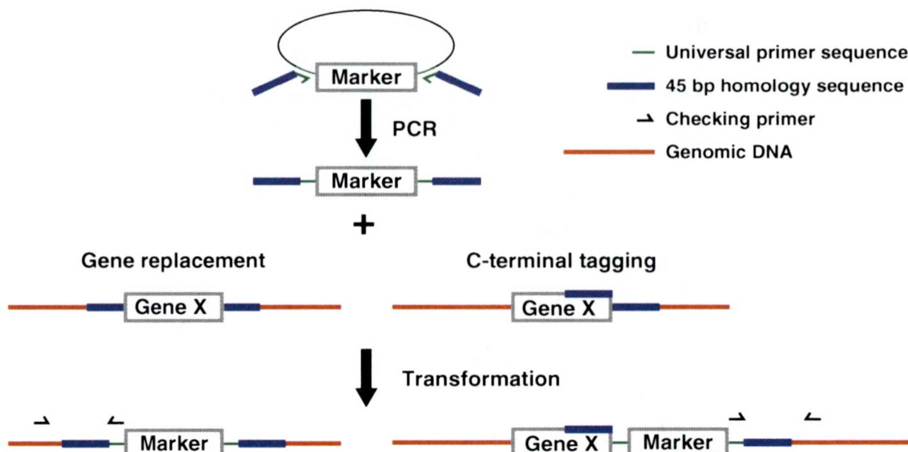

Fig. 1 Schematic of gene replacement and C-terminal tagging. In both cases, the selectable "marker" is PCR amplified off a plasmid using a pair of hybrid primers. The 3′ end of the primers contains sequences specific to the plasmid (*thin green lines*) in a region ideally identical among a family of marker plasmids to achieve flexibility. The 5′ end of the primers is 45 bp homologous to yeast genomic DNA (*thick blue lines*). For gene replacement, the forward homology is to the region immediately 5′ of the start codon of the open reading frame (ORF) of the gene of interest ("Gene X"), while the reverse homology is the reverse complement of the region immediately 3′ of the stop codon of the ORF. For C-terminal tagging, the reverse homology is the same as for gene replacement, while the forward homology is the 45 bp leading up to but not including the stop codon of the ORF. After transformation, checking primers (*thin black arrow lines*) are used in colony PCR to verify proper integration of the DNA fragment. One primer has homology to the plasmid sequence, while the other has homology specific to the region outside of the 45 bp homology used for integration. For C-terminal tagging, the depicted location of checking primers is preferred, as it results in a shorter (and therefore easier to amplify) PCR product

Table 1
Lasers and filter sets used to simultaneously resolve five fluorescent proteins and one fluorescent dye

Laser (nm)	Filter[a]	Fluorophore
405	450/50	CFP
488	505/10 530/30	GFP YFP/citrine
561	590/20 615/25	mOrange mCherry
639	660/16	TO-PRO-3

[a]The number before the slash indicates the center wavelength in nanometers; the number after the slash is the total bandwidth passed by the filter. For example, "450/50" indicates a filter that passes wavelengths from 425 to 475 nm

modified these plasmids to contain *nat* (WSB116) and *hph* (WSB117). We generally do not use auxotrophy (the inability to synthesize an essential metabolite) to mark strains except when both selection and counterselection are required. For example, integration of the loxP–drug–loxP cassette could be carried out in a *ura3* strain so that transformation of a *URA3*-marked plasmid containing the Cre recombinase can be selected for. After induction of Cre expression and removal of the drug marker, the *URA3* plasmid can then be counterselected using 5-FOA, which only allows survival of cells without the plasmid [20]. The uracil auxotrophy may then need to be removed via genetic crosses. Alternatively, Cre expression plasmids containing antibiotic markers are also available [19].

Once the populations have been marked, interactions between populations can be defined. Obviously, the possibilities are practically limitless, and the specifics must be left to the researcher. We chose to base our two-partner cooperative system on complementary nutrient exchange [21].

In yeast, all of the manipulations described above can be achieved through a small set of well-known methods [19]. These include (1) transformation to insert or remove genetic material; (2) colony PCR, a simple and rapid way to check whether transformation was successful; and (3) mating, sporulation, tetrad dissection, and genotyping, which allows genetic features present in two different strains to be recombined into one strain. Below we describe our current protocols for each of these methods.

2 Materials

2.1 Components for Genetic Manipulation

1. Plasmids containing genes encoding fluorescent proteins and/or selectable markers.

2. Antibiotics: 1,000× stock G418 (200 mg/ml), 500x stock hygromycin B (100 mg/ml), 1,000× stock clonNAT (100 mg/ml), 1,000× stock AbA (0.5 mg/ml) (*see* **Note 1**).

3. YPD: 10 g/l Bacto-yeast extract, 20 g/l Bacto-peptone, and bring to volume with diH$_2$O to final 950 ml/l for later glucose supplement. Add 20 g/l Bacto-agar (*see* **Note 2**) if making plates. Add a magnetic stir bar, and autoclave. Using a flame to ensure sterility, add 50 ml 40 % glucose per liter of medium and the appropriate antibiotic, if necessary. Stir to mix. If multiple liters of agar medium are prepared, they may be kept at 50 °C water bath to prevent solidification of the agar.

4. SD: For liquid media, add 6.7 g/l Difco™ yeast nitrogen base (YNB) with ammonium sulfate and without amino acids and 20 g/l glucose, and bring to volume with diH$_2$O. Sterilize using 0.22 μm filter. For plates, add 6.7 g/l YNB, 20 g/l Bacto-agar, and a stir bar, and add diH$_2$O to 950 ml/l. Autoclave. Using sterile technique, add supplements as necessary. Amino acid and nucleobase supplements can be mixed in appropriate proportions [19] in their powder forms in a

sterilized blender and stored at room temperature. Powdered supplements can be weighed and directly added to the media. Finally, add 50 ml 40 % glucose. Use the glucose to wash down any residual supplement powder adhering to the vessel wall and stir.

5. 50 % PEG 2000 (*see* **Note 3**): Dissolve 100 g PEG in 100 ml diH$_2$O. Bring to 200 ml with diH$_2$O. Sterile filter.

6. 1 M LiAc: 102 g/l Lithium acetate dihydrate (102.02 g/mol). Sterile filter.

7. 5 mg/ml Sheared salmon sperm DNA (SS-DNA) [22].

8. Autoclaved water, tubes, and tips.

2.2 Components for Mating, Sporulation, and Tetrad Dissection

1. Sporulation media: 3 g/l potassium acetate, 0.2 g/l raffinose, bring to final volume with diH$_2$O. Autoclave.

2. SCE buffer: 1 M D-sorbitol, 0.1 M sodium citrate (*see* **Note 4**), 60 mM EDTA. Adjust pH to 7.0 with 38 % HCl (*see* **Note 5**). Autoclave using a 20′ sterilization cycle, and remove promptly (*see* **Note 6**).

3. Zymolyase 20T: 30 mg zymolyase 20T dissolved in 10 ml SCE (*see* **Note 7**).

---

3   Methods

3.1 Primer Design and Amplification for Gene Tagging or Replacement

1. Design primers: For gene replacement (Fig. 1), the forward primer should contain 45 base pairs (bp) of homology to the genomic region immediately upstream of the "ATG" codon of the open reading frame (ORF) of the gene to be knocked out, followed by the sequence for the universal forward adapter appropriate for the particular set of plasmids being used. The reverse primer should contain the reverse complement to the 45 bp including and immediately downstream of the stop codon of the target ORF, followed by the reverse complement to the universal reverse adapter appropriate for the plasmid set. For C-terminal tagging (Fig. 1), the reverse primer is designed as in gene replacement, and the forward primer should contain 45 bp homology to the sequence just 5′ of the stop codon of the gene of interest.

2. PCR amplify cassette (antibiotic resistance for gene replacement or fluorescent protein plus selectable marker for C-terminal tagging) off a plasmid (*see* **Note 8**). Check the length of the PCR product using gel electrophoresis.

3.2 Transformation (See Note 9)

1. Inoculate 5 ml YPD culture per transformation and shake at 30 °C until the cell density is between 3×10^6 and 2×10^7 cells/ml (*see* **Note 10**).

2. Prepare a boiling water bath for the SS-DNA. While the water is warming up, harvest the culture in a sterile 50 ml centrifuge tube at $425 \times g$ for 2 min.

3. Pour off the YPD medium, and resuspend the cells in 25 ml of sterile water and centrifuge again.

4. Pour off the water, resuspend the cells in 100 µL 0.1 M LiAc, and transfer the suspension to a 1.5 ml microfuge tube.

5. Pellet the cells at top speed for 15 s, and remove supernatant with a micropipette.

6. Resuspend the cells in about 43 µl of 0.1 M LiAc to a final volume of 50 µl (2×10^9 cells/ml) per transformation.

7. By now the water bath should be boiling. Boil SS-DNA (with cap lock on) for 5 min, and then quickly transfer it to ice (*see* **Note 11**). Vortex briefly to speed up cooling, and then keep on ice.

8. Vortex the cell suspension and pipette 50 µl into 1.5 ml tubes. Pellet the cells at top speed for 15 s, and remove supernatant with a pipette. Vortex to loosen up the pellet.

9. Prepare the transformation mixture (TRAFO) master mix (1.2× the total volume required so that pipetting errors can be accommodated) and keep on ice (*see* **Note 12**): 240 µl 50 % w/v PEG (*see* **Note 3**), 36 µl 1.0 M LiAc, 20 µl 5 mg/ml SS-DNA, and 64-*x* µl sterile dH$_2$O per transformation, where *x* µl is the volume of DNA to be added (*see* **Note 13**). Vortex until completely homogenous (*see* **Note 14**).

10. Add 360-*x* µl TRAFO to each cell pellet (*see* **Note 15**), and mix well by pipetting up and down (*see* **Note 16**). Add *x* µl DNA, and mix well again by vortexing or pipetting.

11. Heat shock cells by placing the tubes in a 42 °C water bath for 40 min (*see* **Note 17**). Mix by inverting every ~10 min.

12. Centrifuge at $3,824 \times g$ for 15 s, and remove TRAFO with a pipette or simply by decanting. Wash cells by resuspending the pellet in 1 ml YPD. Spin again at $3,824 \times g$ for 15 s (*see* **Note 18**). Remove YPD, and pipette 1 ml fresh YPD into each tube. Resuspend the pellet by pipetting it up and down gently.

13. If selection is on complementation of nutrient auxotrophy, cells can be directly plated on selective medium. If selection is on drug resistance, cells need to be incubated for ~2–3 h at 30 °C in 1 ml YPD to express the resistance gene before being plated.

14. To plate, first centrifuge at $3,824 \times g$ for 15 s. Discard 700 µl of the supernatant. Resuspend cells in the remaining 300 µl and plate on 80 % of the surface. Use a sterile toothpick to streak from the plated area to the empty area to maximize the chance of obtaining single colonies.

15. Incubate at 30 °C. Colonies should be visible in 2–3 days.

**3.3 Colony PCR
(See Note 19)**

1. Design primers: One primer should be specific to the cassette used for transformation (for example, the universal primer sequence). The other primer should be outside the 45 bp homology region used for integration (Fig. 1).

2. Using a sterile pipette tip, pick about half of a normal-sized (~1.2 mm diameter) colony without touching the agar beneath it. For S288C, place directly into 15 μl sterile water, vortex, and use 1 μl in a 20 μl PCR (*see* **Note 20**). For RM11, transfer the cells to 15 μl 0.25 % SDS (*see* **Note 21**). Vortex for 30 s, spin at top speed (~17,949 ×*g* in a small centrifuge) for 1 min, and use 1 μl of the supernatant for a 20 μl PCR. This PCR mix must contain a final concentration of 5 % Triton X-100 to neutralize the protein-denaturing SDS [23]. An alternative, and more reliable, method uses LiAc and SDS to lyse cells and requires ethanol precipitation prior to PCR (*see* **Note 19**, ref. 22). Specifically, cells are suspended in 100 μl 200 mM LiAc and 1 % SDS solution and incubated at 70 °C for 15 min. 300 μl 96 % ethanol is added to precipitate DNA. After brief vortexing, DNA is collected by centrifugation at 15,000 ×*g* for 3 min. Precipitated DNA is dissolved in 100 μl TE (10 mM Tris–HCl, 1 mM disodium EDTA, pH 8.0). After spinning cell debris down at 15,000 ×*g* for 1 min, 1 μl supernatant is used for PCR.

**3.4 Mating and
Diploid Isolation**

1. Grow up a small patch of each haploid to be crossed towards the top of a YPD plate (*see* **Note 22**).

2. Use a sterile toothpick to transfer a tiny amount of one strain to an empty region of a YPD plate (*see* **Note 23**), making a small spot (a few millimeters in diameter). With a fresh toothpick, transfer a similarly tiny amount of the other strain to the same spot and mix with the toothpick.

3. Incubate for 3.5 h (S288C) or 2.5 h (RM11) at 30 °C (*see* **Note 24**).

4. Using a toothpick, touch the mixed spot and streak down about a centimeter. Without re-touching the spot, make similar streaks to the left and right of the original streak. This gives three different dilutions of cells on the plate.

5. Use a yeast dissection microscope [19] to isolate diploids. The cytoplasmic "bridge" between mated cells and a small bud at its center is indicative of recently mated diploids.

**3.5 Sporulation,
Tetrad Dissection,
and Genotyping**

1. Pre-sporulate by patching onto YPD and incubating at 30 °C for ~10 h until a thin film of cells is formed (*see* **Note 25**).

2. Inoculate 2 ml sporulation media with about a match head's worth of cells. Incubate at room temperature on a rotator for 2–5 days. Check for tetrads using a light microscope using a 20–40× magnification objective [19].

3. Wash cells with 1 ml sterile water and resuspend in 1 ml sterile water. Store at 4 °C.

4. When ready to dissect, spin down 20 µl of sporulated cells and carefully remove supernatant with a pipette. Add 20 µl SCE, and vortex to resuspend cells. If using RM11, sonicate for 1 min.

5. Add 4 µl zymolyase 20T, and mix by pipetting up and down (*see* **Note 26**). Incubate at room temperature for 20 min (S288C) or 2 h (RM11) (*see* **Note 27**).

6. Gently (to avoid breaking up tetrads) pipette up digested cell suspension and spot onto the top portion of the center strip of a YPD plate. Tilt plate down so that the liquid rolls down the central strip, stopping before the liquid touches the plate wall. Let dry.

7. Use a yeast dissection microscope to separate tetrads into individual spores. Create a grid with a maximum of two tetrads per row [19], one to the left and one to the right of the central strip.

8. Incubate overnight at 30 °C. Use a flame-sterilized scalpel to remove the strip of cells from the middle of the plate (*see* **Note 28**) and return to incubator.

9. Once the spores have germinated and grown into colonies, replica plate onto the appropriate selective media(s). Note that a single genetic locus should, under most circumstances, segregate 2:2 [19]. For instance, two spores will be *MAT*a and two will be *MAT*α. Exceptions can be caused by, for example, gene conversion or traits that are mediated by heritable materials transmissible through the cytoplasm, such as mitochondria.

10. Mating types can be tested using a pair of mating-type testers, one of each mating type. If the entire collection of spores are auxotrophic because of mutations in a set of genes, then the testers should be auxotrophic due to mutation in a different gene to ensure that the resulting diploids are prototrophic. In this case, spread ~200 µl of a saturated culture of each tester strain on its own YPD plate and let dry. Then, replica plate the target strains onto each plate using a sterile velvet. After half a day, replica each plate onto an SD plate. Growth will only occur if the strains are able to mate. If spores are prototrophic, they need to be separately mated to each of the two tester strains, as described above for diploid isolation. Many crosses can be set up on one YPD plate. The mating type of a strain is revealed by the presence or the absence of fused diploids. If this method is necessary, it is better to narrow down the number of strains to be tested based on other markers first.

4   Notes

1. Prepare stocks in deionized (di) H_2O and sterile filter (except for AbA, which should be made in ethanol and needs no sterile filtration) and keep at −20 °C.

2. Not all sources of YNB and agar are appropriate for use with yeast. For example, we have found that YNB and agar obtained from BD biosciences work better than those obtained from USA Scientific.

3. The size of PEG is important. Transformation efficiency for RM11 is very low if PEG 3500 is used. We use PEG3500 to transform S288C, although PEG2000 should work as well.

4. NaH_2PO_4 can be used instead.

5. About 0.5 ml when making a final volume of 500 ml.

6. Ensure that the color of solution did not change during the autoclave process.

7. Freeze down 1 ml aliquots at −20 °C. To use, thaw one tube and make smaller aliquots. Store one at 4 °C for up to 2 months, and freeze the rest for later use.

8. Always use a high-fidelity polymerase for PCR to minimize the possibility of introducing mutations into the amplified fragment. Here is a specific recipe for C-terminally tagging *FBA1* using the pKT plasmids [16] and KOD polymerase (EMD Millipore): 0.5 µl miniprep DNA (~100 ng), 5 µl 10x buffer, 4 µl 25 mM $MgSO_4$, 5 µl 8 mM dNTPs, 0.5 µl primer WSO178 (50 µM), 0.5 µl primer WSO179 (50 µM), 0.5 µl KOD polymerase, 34 µl water (molecular biology grade). WSO178 sequence: 5′-AAGATCACCAAGTCTTTGGAAACTTTCC GTACCACTAACACTTTAggtgacggtgctggtt ta-3′. WSO179 sequence: 5′-GATTCAATACTCATTAAAAA ACTATATCAATTAATTTGAATTAACtcgatgaattcgagctcg-3′. The 45 bp homology sequence is uppercase; the universal primer sequence is lowercase. PCR settings: 94 °C for 2 min, 30 cycles of {94 °C for 30 s, 55 °C for 30 s, and 70 °C for 3 min}, and then 70 °C for 10 min. When *nat* is the template, it is essential to add DMSO to a final concentration of 5 % [11].

9. Transformation is mutagenic. It is therefore best to transform diploids and then sporulate, since one round of meiotic segregation reduces the probability of obtaining an undesired mutation by 50 %. If multiple haploids of the desired genotype derived from the same diploid behave similarly, then background mutations are unlikely to be important. Alternatively, our lab has found that Illumina deep sequencing can reveal the presence of mutations caused by transformation.

10. To ensure that cultures are in exponential phase when it is time to transform, pilot growth experiments may be helpful. Wild-type RM11 grows very rapidly. If the density is $>5 \times 10^7$

cells/ml, dilute to allow the cells to complete at least two divisions in unsaturated conditions. Transformation efficiency remains constant for 3–4 cell divisions.

11. Keep small aliquots of SS-DNA to limit the number of freeze–thaw cycles. Keep on ice when out of the freezer.

12. Keeping the TRAFO master mix on ice is crucial for high efficiency.

13. Use >100 ng DNA. In general, more DNA yields more transformants, but this relationship will likely saturate at some point.

14. This can be visually confirmed by ensuring that no visible "strands" are present in the mixture.

15. TRAFO is very viscous, so pipette slowly to ensure that the correct volume is transferred.

16. Mixing well ensures that the SS-DNA effectively blocks non-specific DNA binding.

17. Adjusting the amount of time for heat shock may be necessary to achieve maximum efficiency. However, 40 min works well for S288C and RM11.

18. Be as gentle as possible at this step, as the cells are very fragile.

19. Colony PCR is quick but can be unreliable. When it fails to work, we have found that a quick DNA extraction before PCR gives reliable results [23].

20. The cell suspension should be turbid.

21. Unlike S288C, boiling of RM11 cells does not provide good template for PCR, although we do not know why. Using detergent effectively lyses the cells and releases their DNA into solution.

22. This can be done for as little as 3 h.

23. The amount should be small enough to not leave a visible film on the plate after transfer.

24. Different strains may require more or less time. Cells should show visible film of growth by this time.

25. Overgrowth will lead to a reduction in sporulation efficiency. However, a suitable number of tetrads should be present even after 12–14 h of pre-sporulation.

26. Vortexing can oxidize the zymolyase.

27. The four spores form a three-dimensional, tetrahedral shape if the ascus wall is undigested. After sufficient digestion, the four spores will have a flat, diamond shape. Underdigestion of the ascus wall will make it difficult to separate the individual spores. Overdigestion will result in tetrads that break apart easily, increasing the chances that four spores in the correct diamond shape are not products of the same meiosis. Overdigestion can also reduce spore viability.

28. Otherwise, growth of this strip will slow down the growth of the haploids nearest to it.

Acknowledgements

Work in the W.S. group is supported by the W. M. Keck Foundation and the National Institutes of Health (Grant 1 DP2OD006498-01).

References

1. Madsen EL (2011) Microorganisms and their roles in fundamental biogeochemical cycles. Curr Opin Biotechnol 22:456–464. doi:10.1016/j.copbio.2011.01.008

2. Clemente JC, Ursell LK, Parfrey LW, Knight R (2012) The impact of the gut microbiota on human health: an integrative view. Cell 148:1258–1270. doi:10.1016/j.cell.2012.01.035

3. Høiby N, Ciofu O, Johansen HK et al (2011) The clinical impact of bacterial biofilms. Int J Oral Sci 3:55–65. doi:10.4248/IJOS11026

4. Zengler K (2009) Central role of the cell in microbial ecology. Microbiol Mol Biol Rev 73:712–729. doi:10.1128/MMBR.00027-09

5. Momeni B, Chen C-C, Hillesland K et al (2011) Using artificial systems to explore the ecology and evolution of symbioses. Cell Mol Life Sci 68:1353–1368. doi:10.1007/s00018-011-0649-y

6. Hagen DC, McCaffrey G, Sprague GF (1986) Evidence the yeast STE3 gene encodes a receptor for the peptide pheromone a factor: gene sequence and implications for the structure of the presumed receptor. Proc Natl Acad Sci 83:1418–1422

7. SGD project Saccharomyces Genome Database. In: SGD. http://yeastgenome.org/cache/genomeSnapshot.html. Accessed 8 Mar 2010

8. Dimitrov LN, Brem RB, Kruglyak L, Gottschling DE (2009) Polymorphisms in multiple genes contribute to the spontaneous mitochondrial genome instability of Saccharomyces cerevisiae S288C Strains. Genetics 183:365–383. doi:10.1534/genetics.109.104497

9. Veatch JR, McMurray MA, Nelson ZW, Gottschling DE (2009) Mitochondrial dysfunction leads to nuclear genome instability via an iron-sulfur cluster defect. Cell 137:1247–1258. doi:10.1016/j.cell.2009.04.014

10. Yvert G, Brem RB, Whittle J et al (2003) Trans-acting regulatory variation in Saccharomyces cerevisiae and the role of transcription factors. Nat Genet 35:57–64. doi:10.1038/ng1222

11. Goldstein AL, McCusker JH (1999) Three new dominant drug resistance cassettes for gene disruption in <I>Saccharomyces cerevisiae</I> Yeast 15:1541–1553. doi:10.1002/(SICI)1097-0061(199910) 15:14<1541 ::AID-YEA476>3.0.CO;2-K

12. Nakazawa N, Iwano K (2004) Efficient selection of hybrids by protoplast fusion using drug resistance markers and reporter genes in Saccharomyces cerevisiae. J Biosci Bioeng 98:353–358. doi:10.1016/S1389-1723(04)00295-6

13. Hentges P, Van Driessche B, Tafforeau L et al (2005) Three novel antibiotic marker cassettes for gene disruption and marker switching in Schizosaccharomyces pombe. Yeast 22:1013–1019. doi:10.1002/yea.1291

14. Day RN, Davidson MW (2009) The fluorescent protein palette: tools for cellular imaging. Chem Soc Rev 38:2887. doi:10.1039/b901966a

15. Sample V, Newman RH, Zhang J (2009) The structure and function of fluorescent proteins. Chem Soc Rev 38:2852. doi:10.1039/b913033k

16. Sheff MA, Thorn KS (2004) Optimized cassettes for fluorescent protein tagging in Saccharomyces cerevisiae. Yeast 21:661–670. doi:10.1002/yea.1130

17. Güldener U, Heck S, Fiedler T et al (1996) A new efficient gene disruption cassette for repeated use in budding yeast. Nucleic Acids Res 24:2519–2524. doi:10.1093/nar/24.13.2519

18. Güeldener U, Heinisch J, Koehler GJ et al (2002) A second set of loxP marker cassettes for cre-mediated multiple gene knockouts in budding yeast. Nucleic Acids Res 30:e23–e23. doi:10.1093/nar/30.6.e23

19. Guthrie C, Fink GR (2002) Guide to yeast genetics and molecular and cell biology, part B, vol 350, 1st edn. Academic, New York

20. Boeke JD, Trueheart J, Natsoulis G, Fink GR (1987) 5-Fluoroorotic acid as a selective agent in yeast molecular genetics. Methods Enzymol 154:164–175

21. Shou W, Ram S, Vilar JM (2007) Synthetic cooperation in engineered yeast populations. Proc Natl Acad Sci U S A 104:1877–1882. doi:10.1073/pnas.0610575104

22. Sambrook J, Russell DW (2001) Molecular cloning: a laboratory manual. Cold Spring Harbor Laboratory Press, New York

23. Lõoke M, Kristjuhan K, Kristjuhan A (2011) Extraction of genomic DNA from yeasts for PCR-based applications. Biotechniques 50:325–328. doi:10.2144/000113672

Chapter 3

Combining Engineering and Evolution to Create Novel Metabolic Mutualisms Between Species

Lon Chubiz, Sarah Douglas, and William Harcombe

Abstract

Synthetic communities can be used as model systems for the molecular examination of species interactions. Manipulating synthetic communities to create novel beneficial interactions provides insight into the mechanisms of cooperation as well as the potential to improve the productivity of industrially relevant systems. Here, we present a general scheme for evolving a mutualism from a bacterial consortium in which one species consumes the by-products of another.

Key words Synthetic ecology, Consortia, Cooperation, Cross-feeding, Evolution

---

1 Introduction

Molecular microbiology has largely been built on the study of monocultures. However, there is growing appreciation that a great deal of microbial life involves interaction between species [1–4]. Moreover, many processes of industrial interest, from degradation of complex molecules to natural product biosynthesis, require multi-species interactions [5–8]. Understanding the mechanistic basis of community behavior therefore represents an exciting new frontier in microbiology. However, the complexity of natural communities makes this a daunting challenge. For this reason, synthetic multi-species consortia represent promising tools for connecting molecular processes to ecological dynamics [2].

A particularly important type of multi-species interaction is metabolic interdependency or cross-feeding. Acquiring metabolites from the excretions of other species is thought to be prevalent in nature, and it has been suggested as one of the reasons that so few microbes can be isolated in laboratory monocultures [2, 9]. Furthermore, there is increasing industrial interest in using cross-feeding to divide the labor of complex metabolic processes such as biofuel production and bioremediation [6, 10]. Several model

Lianhong Sun and Wenying Shou (eds.), *Engineering and Analyzing Multicellular Systems: Methods and Protocols*,
Methods in Molecular Biology, vol. 1151, DOI 10.1007/978-1-4939-0554-6_3, © Springer Science+Business Media New York 2014

cross-feeding systems have been created for laboratory study. For example, Shou et al. constructed an obligate interaction between two auxotrophic yeast strains which provided each other with an essential amino acid and an essential nucleotide precursor, respectively [11]. In another study Hillisland et al. demonstrated that evolution can improve the efficiency of cross-feeding with a syntrophic system in which *Methanococcus maripaludis* consumed waste hydrogen created by *Desulfovibrio vulgaris* [12]. Harcombe combined engineering and evolution to demonstrate the factors necessary for the evolution of costly cooperation between species [13]. He observed that on lactose minimal media *Salmonella enterica serovar* Typhimurium relies on carbon by-products excreted by *Escherichia coli*. Subsequently it was demonstrated that if the *E. coli* was made auxotrophic for an essential amino acid, *S. enterica* could evolve to provide the costly metabolite. This adaptation required both the spatial structure of solid media and by-product provisioning from *E. coli*.

Novel metabolic interactions can be created solely through genetic engineering as demonstrated by the Shou et al. example above [11]. However, this will be difficult in less tractable systems. Here, we build on the methods developed by Harcombe [13] to layout a framework for creating novel mutualisms from cross-feeding association. As discussed previously, the method was developed for an interaction between an *E. coli* auxotrophic for methionine and an *S. enterica*; however, we attempt to broaden the approach for any *E. coli* auxotroph paired with a different species.

2 Materials

All materials are based on the methodology used to create the *E. coli–S. enterica* acetate/methionine cross-feeding consortium, but the methods can be extended to other systems and other metabolites. All media are prepared using standard microbiological sterilization (autoclaving, filtering, or flaming) methods and aseptic technique. Diligently follow all culturing and disposal methods compliant with the biosafety level (BSL) associated with the organisms under investigation. We employ BSL-2 procedures for the species described here.

2.1 Species Selection

2.1.1 Auxotrophic E. coli

Example: *E. coli* K-12 strain JW3910 (Δ*metB* CGSC# 10824, [14]; *see* **Note 1**). The Keio collection contains all single-gene knockouts in *E. coli* and therefore provides numerous different auxotrophs [14]. However, all *E. coli* in the Keio knockout collection are Δ*lacZ* and therefore require the *lacZ* gene to be added back in order to grow on lactose. With the *lac* operon restored, *E. coli* Δ*metB* will metabolize lactose and excrete acetate but will be unable to synthesize methionine (i.e., a Met⁻, Lac⁺ phenotype). Restoration of the *lac* operon can be done via generalized transduction with a P1*vir* lysate grown on a *lac⁺ E. coli* donor as described in detail below.

2.1.2 Species B

General requirements:

1. Species B is able to utilize acetate (but not lactose) as an essential metabolite.

2. Species B can synthesize the compound for which *E. coli* is auxotrophic.

Example: *S. enterica* can synthesize methionine and has no *lac* operon (i.e., a Met⁺, Lac⁻ phenotype) but can metabolize acetate—a by-product of *E. coli* lactose metabolism. Therefore, with lactose as the sole carbon source, *S. enterica* growth is dependent on the growth of *E. coli*.

2.2 Biosynthetic Analogues

Costly compounds such as amino acids, vitamins, and nucleotides are often regulated via end-product inhibition at the level of transcription and/or translation [15–17]. This often results in the inability of microbes to excrete these compounds. Evolving resistance to non-metabolizable or toxic analogues of end products is one way to abolish end-product inhibition, allowing overproduction of the compound. Overproduction can be evolved without first selecting for resistance to a toxic analogue, but the efficiency of selection will be much lower.

Example: Ethionine, a toxic and competitive methionine analog, was used to select for constitutive expression of the *S. enterica* methionine biosynthetic pathway, as described by Lawrence and co-workers [18]. Harcombe found that this initial selection with a biosynthetic analog was insufficient to cause cross-feeding in the *E. coli–S. enterica* consortium; however, the treatment increased the effectiveness of subsequent selection for consortium growth.

2.3 Media and Buffers

A rich (or complete) growth medium that permits each species to grow rapidly, such as Luria Broth (LB), is needed to grow initial cultures of *E. coli*, *S. enterica*, or other organisms of interest.

LB media:

1. Dissolve 10 g tryptone, 5 g yeast extract, and 10 g NaCl in 600 mL deionized water. If making plates, also dissolve 15 g agar.

2. Add water to bring final volume up to 1 L.

3. Adjust pH to 7.0 with 1 M NaOH.

4. Autoclave for at least 15 min and 15 psig.

5. If making plates, let media cool until safe to handle, and then pour ~25 mL per 15 mm petri dish.

Four types of minimal media are needed. We use M9, but any minimal media should work. Add analog stock solution and carbon source to media after autoclaving. Take care to determine the inhibitory concentration of the analog and use a concentration at least twofold above that concentration (*see* Subheading 3.2, **step 2**).

M9 media:

1. Make 5× M9 salts by dissolving the following in 1 L of deionized water and autoclaving for 15 min at 15 psig to sterilize:

   - 64 g $Na_2HPO_4.7H_2O$.
   - 15 g KH_2PO_4.
   - 2.5 g NaCl.
   - 5 g NH_4Cl.

2. To make liquid media, combine 688 mL of deionized, sterile water, 200 mL 5× M9 salts, 2 mL 1 M $MgSO_4$ (sterile), 100 µL $CaCl_2$ (sterile), and 10 mL 10 % [wt/vol] glucose or other carbon sources (sterile).

3. If making solid media plates, make the following two solutions separately. In a 1 L flask, combine 200 mL of 5× M9 salts with 288 mL of deionized water. In a separate 2 L flask, combine 15 g of Bacto agar and 500 mL of deionized water. Autoclave both flasks for 15 min at 15 psig. After autoclaving, add 2 mL 1 M $MgSO_4$ (sterile), 100 µL $CaCl_2$ (sterile), 10 mL 10 % [wt/vol] glucose or other carbon sources (sterile), and desired analog stock solution (optional). Once media has cooled sufficiently for safe handling, pour ~25 mL per 15 mm petri dish.

Analog media:

Contain all nutrients necessary for the growth of species B.

Contain analog at concentration that prohibits growth (*see* Subheading 3.2, **step 2**, **Note 2**).

Example: M9 agar plates with 0.1 % glucose and 6 mM ethionine.

Consortium media:

Contains nutrients that permit growth only if species form mutualism.

Example: M9 agar plates with 0.1 % lactose.

Auxotrophic E. coli selective media:

Contain all nutrients necessary for the growth of auxotrophic *E. coli*.

Example: M9 agar plates with 0.1 % lactose and 40 µg/mL methionine.

Species B selective media:

Contain all nutrients necessary for the growth of species B.

Example: M9 agar plates with 0.1 % glucose.

If *E. coli* from the Keio collection are used, *lacZ* can be inserted with generalized transduction. For the production of phage lysates used in the generalized transduction procedure below, we use the following reagents.

Phage salts (PS) [10 mM $MgSO_4$, 5 mM $CaCl_2$]:

1. To 98 mL of deionized water, add 1 mL of a 1 M $MgSO_4$ stock solution and 1 mL of a 0.5 M $CaCl_2$.

2. Sterilize the solution by passing the entire 100 mL through a 0.22 μm filter.

3. Make 10×10 mL aliquots and store indefinitely at 4 °C.

Luria Broth with 5 mM $CaCl_2$ (LBC):

1. To 10 mL of LB, add 0.1 mL of 0.5 M $CaCl_2$.

2. Sterilize by passing the entire 10 mL through a 0.22 μm filter.

1 M sodium citrate solution:

1. To 50 mL water add 29.4 g of sodium citrate dihydrate.

2. Mix and dissolve the sodium citrate by bringing the final volume to 100 mL.

3. Sterilize by passing the entire 10 mL through a 0.22 μm filter.

---

3 Methods

3.1 Reinsertion of lacZ into an Auxotrophic E. coli via Generalized Transduction

3.1.1 Production of a Donor Phage Lysate

1. Grow a donor Lac⁺ strain such as MG1655 (CGSC# 7740) overnight in 5 mL of LB at 37 °C.

2. Inoculate 3 mL of LBC with 30 μL of the Lac⁺ donor strain and grow with aeration for 1 h at 37 °C.

3. To the donor culture, add 20 μL P1 *vir* phage (*see* **Note 3**).

4. Continue growth with aeration at 37 °C for 4–5 h or until lysis of the culture has been observed (the turbid culture has become transparent or there is visible cell debris).

5. To the donor culture, add 100 μL of chloroform and gently vortex the culture.

6. Centrifuge the culture at $6,000 \times g$ for 5 min to remove cell debris and chloroform.

7. Decant the supernatant into a sealed, sterile glass or plastic tube for long-term storage at 4 °C. This is the donor lysate in which the phage particles will remain viable for several years if stored at 4 °C in the dark.

*3.1.2 Generalized
Transduction*

8. Grow the desired recipient (the strain to receive desired alleles such as the *lac* operon) overnight in 5 mL of LB at 37 °C.

9. Pellet 1 mL of recipient culture by centrifugation at $6,000 \times g$ for 5 min.

10. Resuspend the recipient cells in 0.5 mL of PS.

11. To three separate culture tubes add:

    (a) 0.1 mL of recipient cells in PS + 0.1 mL LBC.

    (b) 0.1 mL of donor lysate + 0.1 mL of PS.

    (c) 0.1 mL of recipient cells + 0.1 mL of donor lysate.

12. Incubate tubes for 20 min at 37 °C.

13. To each tube add 1 mL of LB + 0.1 mL of 1 M sodium citrate solution.

14. Incubate the cultures for 1 h at 37 °C with aeration.

15. Plate the 0.1 mL of each culture onto auxotrophic *E. coli* selective solid medium and grow overnight at 37 °C. Only cells that have successfully acquired the *lac* operon should be able to grow.

16. Restreak colonies on solid selective medium for two transfers to clear any residual bacteriophage.

17. Validate the inheritance of desired alleles by phenotype or genetic techniques such as PCR confirmation.

3.2 Select for Constitutive Expression of the Biosynthetic Pathway in Species B

1. Grow up a colony of species B overnight in LB.

2. Determine the minimum inhibitory concentration of the analog.

    (a) Set up a series of ten flasks each with 10 mL of species B permissive media.

    (b) Dilute the analog twofold in each subsequent flask (*see* **Note 2**).

    (c) Inoculate 10 μL of overnight culture in each flask.

    (d) Incubate the flasks for 2 days at 37 °C.

    (e) Determine the optical density of cells in each flask. The minimum inhibitory concentration is the lowest concentration of analog that allows an optical density of less than half the maximum density that is observed in the gradient.

    (f) Make analog plates with 2× the minimum inhibitory concentration.

3. Wash cells by centrifuging 1 mL of culture at $10,000 \times g$ for 1 min and then resuspending pellet in 1 mL of consortium liquid media.

4. Spread cells evenly on analog plates at a series of densities ranging from 10^6 to 10^8 cells/plate. The range of cell densities is to increase the likelihood of getting colonies when the frequency of resistance mutations is unknown and to ensure that the plate is not too crowded to interfere with selection.

5. Incubation overnight at 37 °C is generally sufficient for *S. enterica* or *E. coli*, but this will vary with species and media.

6. Pick multiple bacterial colonies and re-inoculate colonies on new analog plates, streaking for colonies with a loop or a toothpick.

7. Incubate at 37 °C overnight or until single colonies are visible.

8. Inoculate single colonies in LB liquid medium, grow overnight at 37 °C, and freeze at −80 °C in 20 % [wt/vol] glycerol (*see* **Note 4**).

3.3 Select for Increased Excretion

1. Grow up auxotrophic *E. coli* and analog-resistant species B overnight in their respective liquid permissive media.

2. On consortium media plates mix a lawn of auxotrophic *E. coli* at 10^8 cells/plate and species B at variable concentrations ranging from 10^5 to 10^8 cells/plate (*see* **Note 5**). Plate variable concentrations of species B to account for variable efficiency of evolution of producer phenotype.

3. Incubate to select for increased excretion. 2–3 days are sufficient for *S. enterica* or *E. coli* but longer may be needed for slower growing cells (*see* **Note 6**).

4. If no colonies are visible, transfer to new consortium media plates:

   (a) Add 1 mL of liquid consortium media to plate, and scrub the surface with a sterilized cell spreader.

   (b) Collect liquid from the surface with a pipette, vortex, and spread 100 µL on a new plate.

   (c) Incubate at 37 °C for 2 days.

5. If no colonies appear after first transfer, repeat **step 4** in Subheading 3.3.

6. Pick colonies and re-streak on selective media for each species.

7. Incubate until colonies are visible.

8. Inoculate single colonies of each species in rich liquid medium, grow overnight at 37 °C to saturation, and freeze at −80 °C in 20 % [wt/vol] glycerol.

3.4 Test for Evolved Cross-Feeding by Cross-Streaking

1. Pick a colony of evolved auxotrophic *E. coli*, and streak a single line from left to right on a consortium media plate.

2. Pick a colony of evolved species B and streak perpendicularly across the auxotrophic *E. coli* line in one clean movement. It is important to keep a part of each streaked line as monoculture.

3. Incubate at 37 °C for 1–2 days.

4. Cross-feeding will result in growth where both species overlap and absence of growth where each species is alone on the plate (Fig. 1; *see* **Note 7**).

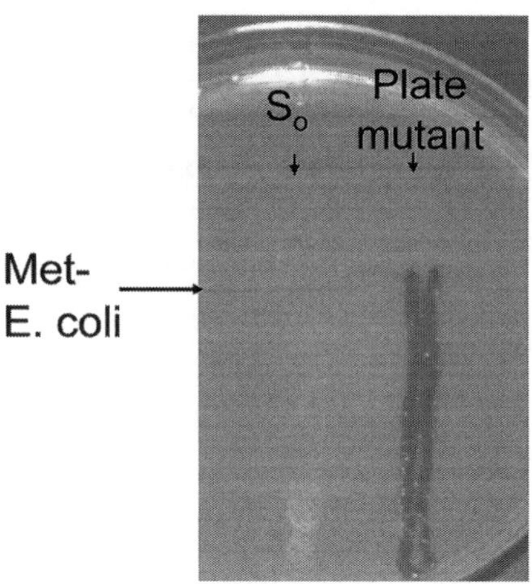

Fig. 1 Cross-streak with *E. coli* and *S. enterica* on a lactose minimal media plate. Auxotrophic *E. coli* was applied in a single line across the middle of the plate with a loop dragging from left to right. Two different *S. enterica* colonies were then streaked from top to bottom. The analog-resistant *S. enterica* (S_0) is on the *left*, while a mutant that arose following plate evolution is on the *right*. No species is able to grow alone as demonstrated by the lack of growth in areas of monoculture (left side of the plate for *E. coli* and top of the plate for *S. enterica*). Growth only occurs in the area where *E. coli* was dragged along with the downward streak of the mutant *S. enterica*. The growth is *blue* because the β-galactosidase (LacZ) indicator 5-bromo-4-chloro-3-indolyl-β-d-galactopyranoside (X-Gal) was added to the plate at 40 μg/mL

4   Notes

1. Strain reference numbers are for the Coli Genetic Stock Center (CGSC), Yale University, http://cgsc.biology.yale.edu.

2. If the analog will not dissolve to create an aqueous stock solution, dissolve in a biologically tolerant solvent such as ethanol, dimethyl sulfoxide, or *N,N*-dimethylformamide. Alternatively, in the case of organic acids and bases (like ethionine), aqueous solutions can be obtained by slowly titrating the compound into solution with 1 M NaOH or 1 M HCl.

3. To avoid transducing unwanted mutations into the auxotrophic *E. coli* the phage should have been previously propagated on a wild-type background such as MG1655.

4. In some cases selection for analog resistance may be sufficient to establish novel resource provisioning. It is worth using the cross-streaking test (Subheading 3.4) to determine isolate phenotype.

5. It is important to do this selection on plates as opposed to liquid media. As demonstrated in Harcombe [13] spatial structure increases selection for costly resource provisioning.

6. If dense lawns form, this may mean that analog-resistant species B produces the compound of interest and selection is not necessary (*see* **Note 4**).

7. If growth occurs where either species is alone then it demonstrates that growth is not the result of novel cross-feeding. An alternative evolutionary strategy is for each species to become self-sufficient. This is particularly likely in the auxotrophic *E. coli* as loss of some biosynthetic enzymes can be compensated for by other proteins. Additional enzymes in a biosynthetic pathway can be removed if this is a problem.

References

1. West SA, Griffin AS, Gardner A et al (2006) Social evolution theory for microorganisms. Nat Rev Microbiol 4:597–607

2. Wintermute EH, Silver PA (2010) Dynamics in the mixed microbial concourse. Genes Dev 24:2603–2614

3. Marx CJ (2009) Microbiology. Getting in touch with your friends. Science 324:1150–1151

4. Schink B (2002) Synergistic interactions in the microbial world. Antonie Van Leeuwenhoek 81:257–261

5. Pettit RK (2009) Mixed fermentation for natural product drug discovery. Appl Microbiol Biotechnol 83:19–25

6. Zuroff TR, Curtis WR (2012) Developing symbiotic consortia for lignocellulosic biofuel production. Appl Microbiol Biotechnol 93:1423–1435

7. McInerney MJ, Struchtemeyer CG, Sieber J et al (2008) Physiology, ecology, phylogeny, and genomics of microorganisms capable of syntrophic metabolism. Ann N Y Acad Sci 1125:58–72

8. Smid EJ, Lacroix C (2013) Microbe-microbe interactions in mixed culture food fermentations. Curr Opin Biotechnol 24:148–154

9. Vartoukian SR, Palmer RM, Wade WG (2010) Strategies for culture of "unculturable" bacteria. FEMS Microbiol Lett 309:1–7

10. Brune KD, Bayer TS (2012) Engineering microbial consortia to enhance biomining and bioremediation. Front Microbiol 3:203

11. Shou W, Ram S, Vilar JMG (2007) Synthetic cooperation in engineered yeast populations. Proc Natl Acad Sci U S A 104:1877–1882

12. Hillesland KL, Stahl DA (2010) Rapid evolution of stability and productivity at the origin of a microbial mutualism. Proc Natl Acad Sci U S A 107:2124–2129

13. Harcombe W (2010) Novel cooperation experimentally evolved between species. Evolution 64:2166–2172

14. Baba T, Ara T, Hasegawa M et al (2006) Construction of *Escherichia coli* K-12 in-frame, single-gene knockout mutants: the Keio collection. Mol Syst Biol 2:2006.0008

15. Umbarger HE (1969) Regulation of amino acid metabolism. Annu Rev Biochem 38:323–370

16. Ishii K, Shiio I (1968) Regulation of purine ribonucleotide synthesis by end product inhibition. I. Effect of purine nucleotides on inosine-5′-phosphate dehydrogenase, xanthosine-5′-phosphate aminase and adenylosuccinate lyase of *Bacillus subtilis*. J Biochem 63:661–669

17. Nierlich DP, Magasanik B (1965) Regulation of purine ribonucleotide synthesis by end product inhibition. The effect of adenine and guanine ribonucleotides on the 5′-phosphoribosyl-pyrophosphate amidotransferase of *Aerobacter aerogenes*. J Biol Chem 240:358–365

18. Lawrence DA, Smith DA, Rowbury RJ (1968) Regulation of methionine synthesis in *Salmonella typhimurium*: mutants resistant to inhibition by analogues of methionine. Genetics 58:473–492

Chapter 4

Design, Construction, and Characterization Methodologies for Synthetic Microbial Consortia

Hans C. Bernstein and Ross P. Carlson

Abstract

Engineered microbial consortia are of growing interest to a range of scientists including bioprocess engineers, systems biologists, and microbiologists because of their ability to simultaneously optimize multiple tasks, to test fundamental systems science, and to understand the microbial ecology of environments like chronic wounds. Metabolic engineering, synthetic biology, and microbial ecology provide a sound scientific basis for designing, building, and analyzing consortium-based microbial platforms.

This chapter outlines strategies and protocols useful for (1) in silico network design, (2) experimental strain construction, (3) consortia culturing including biofilm growth methods, and (4) physiological characterization of consortia. The laboratory and computational methods given here may be adapted for synthesis and characterization of other engineered consortia designs.

Key words Synthetic consortia, Microbial communities, Elementary flux mode analysis, P1 phage transduction, Chemostat, Biofilm, Microsensor, *Escherichia coli*

1 Introduction

Microbes do not typically exist as monocultures in nature; instead, most microbial processes are consortial. Synthetic consortia with engineered division of labor are being utilized to explore fundamental concepts associated with different types of microbial interactions [1, 2]. Many synthetic consortia are built around a naturally occurring strategy of syntrophy, a category of microbial interactions that is mutually beneficial [3–8]. This chapter illustrates consortia-focused methodologies for design, engineering, and analysis based on a previously described synthetic consortium composed of *Escherichia coli* deletion mutants [3]. The engineered coculture utilizes a nature-inspired motif of strain cross-feeding which utilizes positive feedback to encourage mutualistic interactions. The consortium is composed of a glucose-positive *E. coli* "producer strain" and a specialized by-product consuming *E. coli*

Lianhong Sun and Wenying Shou (eds.), *Engineering and Analyzing Multicellular Systems: Methods and Protocols*,
Methods in Molecular Biology, vol. 1151, DOI 10.1007/978-1-4939-0554-6_4, © Springer Science+Business Media New York 2014

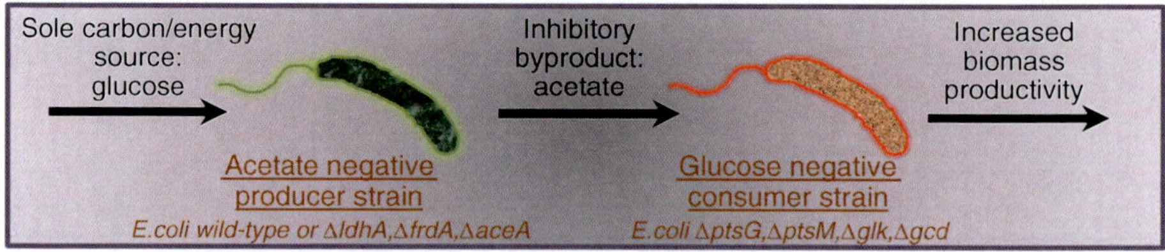

Fig. 1 Schematic diagram illustrating the engineered coculture described in Bernstein et al. 2012. The glucose positive strain's by-products feed the scavenger strain. Engineered *E. coli* serve as a convenient host system for synthetic consortia due to their metabolic flexibility, extensive literature base, and well-established genetic systems

"scavenger strain." The functionality of this system differs from a wild-type *E. coli* monoculture system by enabling simultaneous consumption of both glucose and inhibitory by-products such as acetate. Figure 1 illustrates the example engineered community metabolite exchange theme. Wild-type *E. coli*, in the presence of glucose, will preferentially consume glucose before catabolizing by-products such as acetate due to catabolite repression [9]. Acetate is an inhibitory metabolite that negatively influences growth and lowers process yields due to a combination of toxicity and a diversion of substrate carbon away from desired products. Acetate accumulation represents a significant problem in large-scale bioprocess applications. The scavenger strain relieves system organic acid inhibition and captures substrate carbon that would otherwise be lost.

This chapter describes approaches useful for constructing and analyzing synthetic consortia. The chapter topics include (1) in silico metabolic network analysis, (2) experimental strain construction via gene deletions, (3) consortia culturing including chemostat and biofilm growth methods, and (4) physiological characterization of consortia using HPLC, microscopy, and microelectrodes. The methods presented can be used as a starting point to synthesize and characterize a wide variety of synthetic consortia.

2   Materials

2.1   *In Silico, Stoichiometry-Based Metabolic Pathway Analysis*

1. Stoichiometric modeling software: Numerous software platforms exist for performing elementary flux mode analysis including Cell Net Analyzer (CNA; Max Planck Innovation GmbH) [10], Elementary Flux Mode Tool (EFMTool; ETH-CSB Zurich) [11, 12], and METATOOL [13]. Additional useful systems biology tools like COBRA (http://opencobra.sourceforge.net) and CoPasi (www.copasi.org) are also available.

2. Matrix manipulation software: In silico network analysis output is succinctly represented as matrices. Efficient matrix

manipulation software packages such as MATLAB or open-source GNU octave are recommended. For modestly sized models, spreadsheet applications like MS Excel are sufficient.

3. Stoichiometric model: Obtain or build a stoichiometric reaction model for central carbon metabolism of interest [14, 15]. In addition to manuscripts and book chapters, public model repositories contain numerous microbial models (e.g., www.ebi.ac.uk/biomodels-main/) (*see* **Note 1**).

2.2 P1 Phage Transduction and Antibiotic Resistance Cassette Curing

1. Antibiotic stock solutions (1,000× running concentration): 100 mg/ml kanamycin, 100 mg/ml chlortetracycline, and 100 mg/ml ampicillin.

2. Luria-Bertani (low-salt LB) medium variants: LB supplemented with 25 mM Na_3 citrate; LB containing 200 μg/ml heat-inactivated (autoclaved) chlorotetracycline and 100 μg/ml ampicillin; LB agar (14 g/l agarose) containing 25 mM Na_3 citrate and 100 μg/ml kanamycin; LB agar containing 100 μg/ml ampicillin; Z-broth: LB containing 5 mM $CaCl_2$ (1 M $CaCl_2$ stock solution autoclaved separately and added to LB after autoclaving).

3. Citrate buffer: 0.1 M citrate buffer pH 5.5.

4. P1 phage lysate (from *E. coli* MG1655 wild type or desired *E. coli* gene deletion strain) and chloroform.

5. *E. coli* Keio gene deletion collection (Thermo Scientific) or desired deletion mutants containing an FRT-flanked kanamycin cassette (or appropriate antibiotic selection marker) gene replacement [16, 17]. Temperature-sensitive *pFTA* plasmid [18], electroporator (or materials for chemical transformation), and SOC medium (Life Technologies, CA). Appropriate PCR primers which flank desired gene deletion(s) (*see* **Note 2**).

2.3 Culturing and Fermentation

1. Culturing medium: M9 minimal media, pH = 6.8–7.0, 6 g/l $NaHPO_4$, 3 g/l KH_2PO_4, 1 g/l NH_4Cl, 0.5 g/l NaCl, 1 ml/l of 1 M $MgSO_4 \cdot 6H_2O$, and 10 ml/l of trace metal stock solution (0.55 g/l $CaCl_2$, 0.1 g/l $MnCl_2 \cdot 4H_2O$, 0.17 g/l $ZnCl_2$, 0.043 g/l $CoCl_2 \cdot 6H_2O$, 0.06 g/l $Na_2MoO_4 \cdot 2H_2O$, 0.06 g/l $Fe(NH_4)_2 (SO_4)_2 \cdot 6H_2O$, 0.2 g/l $FeCl_3 \cdot 6H_2O$) [19]. Carbon sources should be added to final concentrations of 4 and 2 g/l for glucose and sodium acetate, respectively. Stock solutions of carbon sources, 1 M $MgSO_4 \cdot 6H_2O$, and trace metals should be autoclaved separately. M9 solid agar petri dishes containing appropriate carbon source can be prepared by adding 14 g/l agarose to basic salt solution prior to autoclaving. Additional solutions are added after autoclaving as with M9 broth.

2. Reactors for batch and continuous culturing: 1 l glass vessel (straight-lipped beaker) equipped with aseptic sampling

port(s), gas diffuser, magnetic stir bar agitator, flow inlet port (for continuous operation), and a dipstick outlet port (for continuous operation) (*see* **Note 3**). 250 ml Erlenmeyer shake flasks for preliminary, screening batch growth experiments.

3. Colony biofilms: M9 agar plates (1 % carbon source w/v), sterile 0.22 μm pore 25 mm diameter polycarbonate membranes (GE Water and Process Technologies, K02BP02500), sterile forceps, sterile phosphate buffer saline solution (PBS; pH 7.4), tissue homogenizer (e.g., Polyscience Tissue Homogenizer model K-120; Polysciences), and sterile homogenizing blade(s).

2.4 HPLC Analysis

1. Agilent 1200 series HPLC (or equivalent) equipped with a variable wavelength and refractive index detector (VWD and RID, respectively).

2. Column and mobile phase: HPX-87H column (BioRad) and filtered (0.45 μm) 0.005 M H_2SO_4 in DI or nanopure water for mobile phase.

3. Standards and dilution matrix: All standards should be made in M9 media with no added carbon source or $MgSO_4$. The primary, multicomponent standard solution used for calibration curve(s) contains 5 g/l glucose and 3 g/l of the following, lactate, formate, succinate, acetate, and fumarate (optional), as well as 3 g/l ethanol (optional) (*see* **Note 4**). Acidifying dilution matrix: a 2× mobile-phase solution (0.01 M H_2SO_4 in DI or nanopure) containing 5 g/l fucose as the internal standard.

2.5 Microscopy and Imaging Analysis

1. Microscope and image analysis software: An epi-fluorescent microscope (Nikon Eclipse E-800 or equivalent) equipped with standard FITC and TRITC filters and appropriate objectives. Imaris (Bitplane) and Metamorph (Molecular Devices) image analysis software (or equivalent) (*see* **Note 5**).

2. Cryosectioning: A Leica CM 1850 cryostat (or equivalent that can be operated at −20 °C), TissueTek O.C.T. tissue embedding medium, dry ice, and razor blade(s). Thin, rectangular stainless steel coupon(s) typically 25×75×1 mm. Positively charged microscope slides (VWR Superfrost Plus; 25×75×1 mm).

3. Fluorescent reporter protein plasmids: Plasmids such as pRSET-mcitrine and pRSET-td-tomato (Dr. Tsien, UC, San Diego) or equivalents that constitutively express fluorescent yellow–green protein and red fluorescent protein.

2.6 Oxygen Analysis by Microsensor

1. Oxygen microsensors (microelectrodes): Clark-type oxygen sensors equipped with guard cathode designed with ≤5-s response time and <5 % stirring sensitivity (OX microsensors, Unisense A/S, DK). Tip diameter(s) should be 10–50 μm depending on the required spatial accuracy.

2. Amplification hardware and data acquisition software: Unisense multimeter (or equivalent amplifier) coupled with SensorTrace (Unisense A/S, DK) data collection software.

3. Positioning equipment: Standard micromanipulator with manual XYZ positioning (at least mm scale accuracy), motorized stage (at least Z-direction with µm scale accuracy), mounting stage, and motor controller (must be compatible with positioning software).

4. Calibration standards: Anoxic standard prepared from DI water containing 100 mM NaOH and 100 mM ascorbic acid. The oxic standard can be prepared from air-saturated DI water (sparged with 1 l per minute (LPM) air) (*see* **Note 16**).

3   Methods

3.1   *In Silico Design and Testing*

Engineered microbial communities can be modeled in silico prior to synthesis and characterization. In silico analysis can provide a rational basis for identifying gene deletion or over-expression targets that encourage or enforce interactions between strains. The authors typically utilize the unbiased, stoichiometric network analysis approach known as elementary flux mode analysis (EFMA) to construct in silico metabolic representations of relevant metabolisms [5, 20]. EFMA defines a network's metabolic potential based on a complete listing of the simplest, non-divisible biochemical pathways (flux distributions). Using the simplest pathways as building blocks, it is possible to perform a "bottom-up" systems biology analysis of individual microbe and community metabolisms. Stoichiometric models do not require extensive enzyme parameter sets like most kinetic models.

1. Stoichiometric models typically start from annotated genome databases and literature reviews. The size and scope of the models can vary depending on research goals. The central metabolism is often sufficient to represent basic physiological characteristics although larger models termed "genome scale" are growing in popularity. The authors recommend software like CNA [10] for building and testing stoichiometric models.

2. Stoichiometric modeling methods, like EFMA, typically assume a steady state for intracellular metabolites. Intracellular metabolite synthesis and consumption are balanced via user-defined, exchange metabolites which represent potential physiological substrates and by-products. Typical substrates would be electron donors and acceptors such as glucose and oxygen, respectively, and typical by-products would be organic acids such as acetic acid. Designation of metabolic sources and sinks plays a critical role in model output and must be considered carefully.

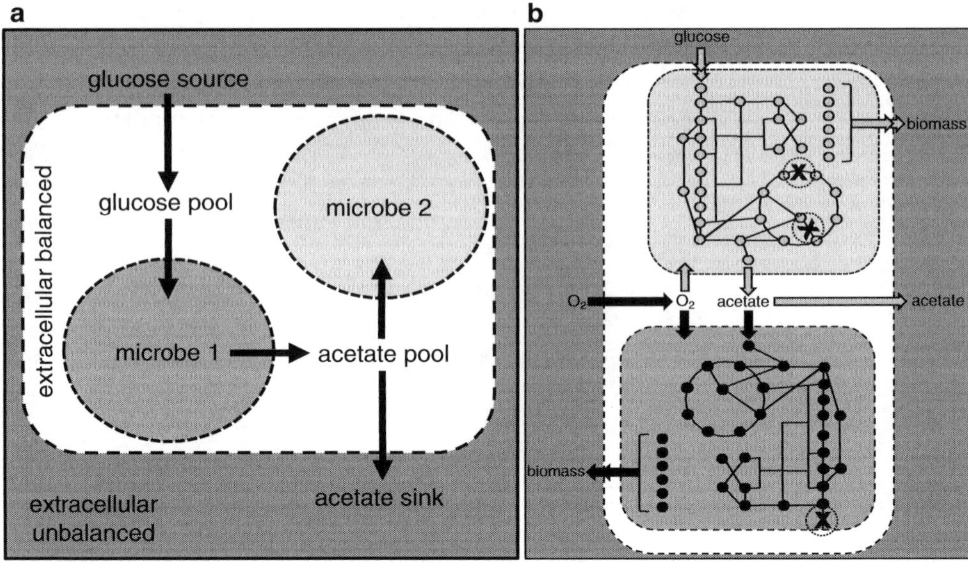

Fig. 2 (**a**) Illustration of four different mass-balanced compartments in a two-microbe community elementary flux mode (cEFM). (**b**) Illustration of an interacting cEFM model based on a metabolic design analogous to systems studied in Bernstein et al. 2012. The *light gray* microbe has gene knockouts in *frdA* and *aceA*, while the *dark gray* organism has gene knockouts associated with glucose uptake (*ptsG, ptsM, glk*) which are indicated with a single "X"

3. To create interacting consortia models, individual microbe models are constructed with each organism's metabolites and reactions partitioned to the appropriate mass-balanced compartment. Individual microbe models are linked through an extracellular mass-balanced, exchange space where metabolites are permitted to be transported from one organism to another. For example, a binary consortium model would possess four independently mass-balanced compartments: microorganism 1, microorganism 2, a balanced extracellular exchange compartment, and the unbalanced extracellular source/sink compartment (*see* Fig. 2). For additional consortia modeling approaches see Taffs et al. [5].

4. Consortium metabolic models are analyzed for mathematically defined metabolic pathways (elementary flux modes) using software highlighted in Subheading 2.1. The output can be succinctly represented as a mathematical matrix where the rows represent distinct elementary flux modes, columns represent model reactions, and matrix elements represent relative fluxes through each model reaction. Zero elements indicate that the model reaction is not utilized by the particular elementary flux mode.

5. EFMA identifies all biologically relevant, mathematically unique combinations of reactions that produce a mass-balanced

metabolic pathway (flux distribution). By systematically sorting these pathways using matrix manipulations, it is possible to identify which reactions are or are not required for a desired consortium functionality. By knowing which reactions are not required, it is possible to identify rational targets for gene deletions (*see* Fig. 2). Examples of output analysis for gene deletion targets can be found in [14, 21].

3.2 Construction of Deletion Mutants

The following section details a method for creating gene deletions in *E. coli* using a P1 phage transduction technique and the Keio *E. coli* mutant library. This method has been developed by combining previously reported transduction techniques [16–18]. These steps may be repeated to generate mutants with multiple gene deletions. Figure 3 illustrates the mechanistic steps taking place during a P1 phage transduction.

1. Add 100 µl of stock P1 phage lysate (approximately 10^8–10^9 pfu/ml) to 10 ml of an overnight donor strain grown in Z-broth and diluted to $OD_{600} = 0.5$. The donor strain is typically a Keio knockout collection strain containing a kanamycin resistance cassette. Create a control, donor strain culture with no added P1 phage.

2. Incubate the two cultures (donor strain + P1, control) for 4–6 h at 37 °C until visually obvious cell lysis has occurred in the $P1^+$ culture as compared to control culture. Alternatively, the $P1^+$ culture OD_{600} can be followed with time (*see* **Note 6**).

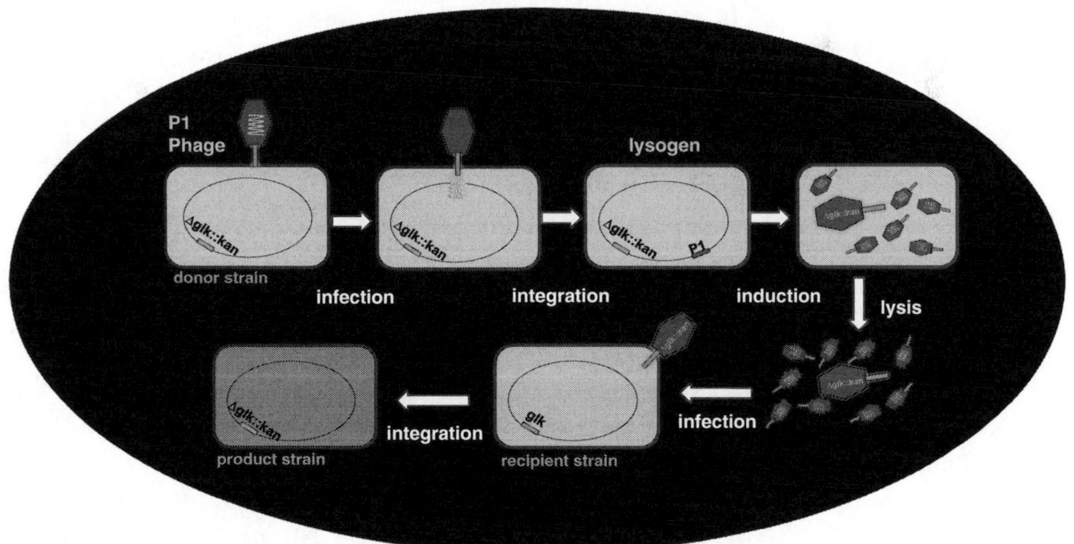

Fig. 3 Schematic drawing illustrating the mechanistic steps of P1 phage transduction in *E. coli*. This illustration uses the single deletion of glucose kinase (*glk*) as an example. Modified from a schematic in Medchrome: Medical and Health Articles

3. Centrifuge lysed P1$^+$ culture (2,000 g, 10 min), and transfer supernatant into sterile tubes (1.5 ml Eppendorf tube or equivalent).

4. Add 100 µl of chloroform to supernatant, vortex, and store at 4 °C. P1 phages are suspended in the aqueous phase.

5. Concentrate 6 ml of a fresh overnight recipient strain culture (target strain to receive the gene deletion) grown in Z-broth by centrifugation (2,000 g, 10 min).

6. Resuspend cell pellet in 4 ml of fresh Z-broth.

7. Prechill and maintain citrate buffer on ice (0–4 °C).

8. Inoculate concentrated recipient strain with 100 µl of aqueous-phase P1 phage solution (*from* Subheading 3.2, **step 4**) and incubate at 37 °C for 20 min with gentle inversions to facilitate phage adsorption.

9. Add 8 ml of cold citrate buffer, and pellet cells via centrifugation (2,000 g, 10 min).

10. Remove unadsorbed phage by washing and centrifuging twice with 10 ml of cold citrate buffer.

11. Resuspend washed cells in 10 ml of LB supplemented with 25 mM Na$_3$ citrate and incubate for 60 min in a shaker at 37 °C (*see* **Note 7**).

12. Recover cells by centrifugation (2,000 g, 10 min), and repeat cold citrate buffer wash twice (*see* Subheading 3.2, **step 4**).

13. Resuspend washed cells in 300 µl citrate buffer, and plate 100–300 µl of cell suspension on LB agar supplemented with 25 mM Na$_3$ citrate and 100 µg/ml kanamycin (or appropriate selective antibiotics). Incubate agar plates overnight at 37 °C. This step will isolate the desired *E. coli* deletion mutant containing the kanamycin cassette resistance marker in place of the targeted gene. Store strain in a cryogenic vial with 20 % sterile glycerol at –80 °C.

14. Begin kanamycin resistance cassette curing process by transforming the product strain from Subheading 3.2, **step 13**, with *pFTA* plasmid using either electroporation or chemical transformation. Immediately incubate transformant culture in SOC medium (~3 ml SOC per 100 µl transformant culture; for electroporation) at 30 °C for 30 min (*see* **Notes 7** and **8**). Plate 100 µl aliquot of transformed culture on LB agar + 100 µg/ml ampicillin and incubate at 30 °C.

15. Inoculate 5 ml of LB containing 200 µg/ml heat-inactivated (autoclaved) chlortetracycline and 100 µg/ml ampicillin with *pFTA*-transformed colonies from previous step. Incubate overnight (5–10 h) at 30 °C. This step will cure the chromosomally integrated kanamycin resistance gene by inducing expression of the *pFTA* flippase gene.

16. Spread-plate multiple volumes (from 5 to 100 µl) of overnight culture from previous step on LB agar (no antibiotics) to ensure the growth of individual colonies. Incubate at 37 °C (*see* **Note 8**).

17. Screen 8–10 colonies from previous step for loss of kanamycin resistance cassette by replica plating colonies on LB agar with and without 100 µg/ml kanamycin, and incubate plates at 37 °C. Successfully cured strains will not grow on LB agar plates containing kanamycin.

18. Verify successful chromosomal gene deletions via PCR with appropriate primers. Compare the size of PCR product against both the *E. coli* wild-type control and the recombinant strain containing the kanamycin cassette (product from Subheading 3.2, **step 13**) (*see* **Note 9**). Store PCR-verified product in a cryogenic vial with 20 % sterile glycerol at −80 °C.

3.3 Batch and Continuous Culturing

This section describes a method used previously to characterize synthetic *E. coli* binary cultures [3]. Achieving chemostasis in continuous culture while maintaining two (or more) stable microbial populations can be challenging and typically requires a priori knowledge of growth parameters such as the maximum specific growth rates for each community member. Here, the authors recommend performing batch characterization prior to continuous growth experiments. The bioreactor culturing method is based on a reactor design that can be assembled in the laboratory (*see* **Note 3** and Fig. 4). Shake flasks are also convenient for initial culture characterization (*see* **Note 10**).

1. Prepare liquid inocula from fresh overnight cultures grown in M9 plus appropriate carbon source (i.e., strain-specific carbon sources such as glucose or acetate). Bring fresh overnight cultures into exponential growth phase by transferring 1–3 ml to fresh media, and follow growth with OD_{600} (*see* **Note 10**). Collect exponentially growing cells by centrifugation (2,000 g, 10 min). Resuspend in experimental feed media (*example*: when chemostat feed contains M9 + 4 g/l glucose, centrifuge *glucose-negative* mutant cells cultured on acetate medium, decant medium, and resuspend in M9 + 4 g/l glucose). Dilute all cultures to an OD_{600} of 0.1 in experimental feed medium (*see* **Note 13**). Mix culture inocula (typically in a 1:1 volume ratio for binary cultures). 5–12 ml of $OD_{600} = 0.1$ culture is typically used to inoculate reactor volumes of 300 ml.

2. Concurrently, fill sterile reactor vessel with 300 ml sterile M9 (containing appropriate carbon source; *see* Subheading 2.3). Save 5 ml of feed medium as a sterile control for downstream analysis (*see* Subheading 3.4). Inoculate reactors with binary strain mixture.

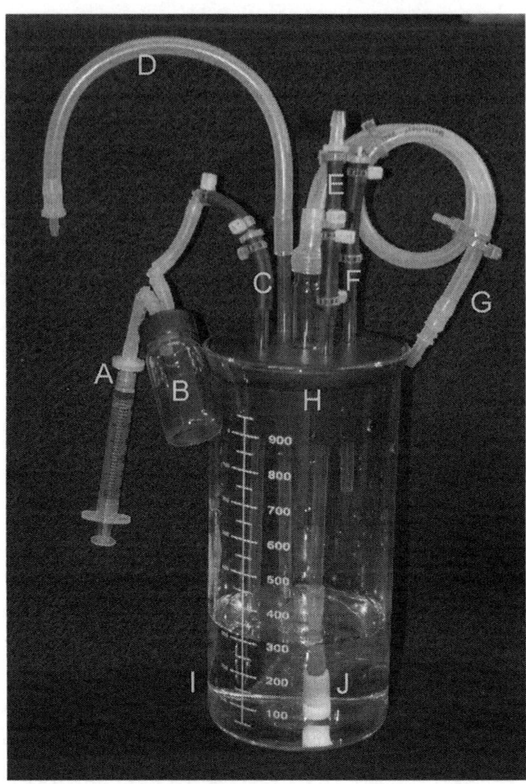

Fig. 4 Photograph of reactor system which can be assembled in the laboratory. (**a**) 5 ml syringe connected via luer lock to a 0.45 μm liquid filter. The in-line liquid filter may be sterilized with the reactor. (**b**) 20 ml collection vial with gasket sealed lid which connects (airtight) to the syringe and sample port. (**c**) Sample port extended into culture liquid volume via glass rod dipstick. (**d**) Out-channel port connected to adjustable glass dipstick which sets reactor liquid volume during continuous operation. (**e**) Aeration or sparge port connected to glass dipstick coupled with porous air stone submerged in culture liquid volume. (**f**) In-channel port connected to glass dipstick held above culture liquid volume. (**g**) Off-gas port coupled with tube coil to prevent airborne contamination. (**h**) Rubber stopper (appropriate size for 1 l cylindrical flask) with bored holes to accommodate glass tube inserts. (**i**) 1 l cylindrical glass flask. (**j**) Porous air stone positioned above magnetic stir bar

3. Incubate inoculated reactor at 37 °C while agitating with a magnetic stir bar (150 rpm). Start air sparge immediately (*see* **Note 11**). Chemostat operation is typically preceded by a batch-phase incubation long enough for the culture to obtain a cell density comparable to the known/expected steady-state value (typically 5–24 h depending on the carbon source and respective growth rates).

4. To start chemostat operation, initiate medium feed at a volumetric flow rate appropriate for the desired dilution rate (dilution rate (D) = volumetric medium flow rate/reactor culture volume) (*see* **Note 10**). Reactor volume must be kept constant; this is achieved with an adjustable *dipstick* connected to the outflow pump (*see* **Note 3** and Fig. 4).

5. Aseptically sample reactor at regular time intervals (typically every 1–2 doubling times). Collect a sample volume large enough for OD_{600}, pH, and HPLC measurements (typically 3–5 ml; *see* **Note 12**). Some applications may also require additional sample preparation for qPCR, microscopy, etc. An operational definition of when reactor chemostasis is achieved can be defined as when measurements stay within ±5 % of the *steady-state* mean values for ≥3 residence times (residence time = 1/D).

6. At the conclusion of chemostat operation, aseptically collect the entire culture volume. A cell dry weight vs. OD_{600} calibration curve can be constructed by dividing the reactor culture into four portions and diluting each volume with sterile experimental feed medium to obtain four different OD_{600} values. Record OD_{600} values, aliquot 50 ml to separate centrifuge tubes, and collect cell pellets via centrifugation (2,000 g, 20 min). To remove salts, wash cell pellets twice with centrifugation in sterile DI water at room temperature. Resuspend washed cells in 5 ml of sterile DI water. Transfer resuspended pellets to clean, pre-weighed 10 ml glass test tubes. It is often necessary to rinse the tube which contained the cell pellet again (once or twice with 2 ml) to successfully transfer all biomass.

7. Incubate glass test tubes (containing 5–9 ml of cell suspension in DI water) in drying oven at 95–99 °C until dry, typically for 24 h. Allow glass test tubes to cool to room temperature. Weigh tubes containing dried cells, and subtract the pre-weighed values. Use these data in combination with their respective OD_{600} values for constructing a cell dry weight vs. optical density standard curve for each reactor condition (*see* **Note 13**). Other methods may be used to determine the relative abundance of each community member (*see* **Note 14**).

3.4 Extracellular Metabolite Analysis (HPLC)

Comprehensive characterization of microbial consortial behavior requires the quantitative assessment of external metabolites including substrates and by-products. High-performance liquid chromatography (HPLC) is an appropriate tool for assaying many common substrates and secreted metabolites. Other methods such as gas chromatography or colorimetric assays may also be appropriate. Here, the authors present a brief description of an HPLC method used to simultaneously quantify analytes: glucose, lactate, succinate, formate, fumarate, acetate, and ethanol.

1. Pre-filter all chemostat and calibration samples with a 0.45 µm syringe filter to remove particulates.

2. Dilute calibration standard (*see* Subheading 2.4, **item 3**) in blank media (M9 with no carbon source) to a minimum of

four different concentrations (*see* **Note 4**). Avoid setting concentrations below or near HPLC detection limits.

3. Mix samples in a 1:1 ratio (v/v) with acidifying dilution matrix (*see* Subheading 2.4, **item 3**) in HPLC vials. Ensure enough sample volume for triplicate injections.

4. HPLC method (Agilent 1200 series HPLC coupled with a Bio-Rad HPX-87H column): (a) set mobile-phase flow rate to 0.6 ml/min, (b) set injection volume to 20 µl, (c) set column and detector temperature to 25 to 50 °C (*see* **Note 4**), (d) set VWD to 210 nm, (e) ensure that RID is flushed with mobile phase, and (f) set individual injection run time to 20 % more than the retention time for the last analyte measured.

5. Inject experimental samples and standards in triplicate, and ensure that calibration standards are ordered from least to highest concentrations to reduce potential sample carryover. Process chromatograms by normalizing the integrated area associated with each analyte by the corresponding signal for the internal standard (typically fucose measured on the RID).

6. Net culture-specific rate for production of by-product i can be calculated using the following relationship: $q_i = D \times (C_{i,\text{out}} - C_{i,\text{in}})/(x)$ where D is the dilution rate (h^{-1}), C_i is the concentration of by-product i leaving (out) and entering (in) the reactor, and x is the biomass concentration (g dry biomass/l). Culture yields can be defined using two different methods. Biomass yield on glucose can be calculated as $Y_{x/\text{glc}} = (x_{\text{out}} - x_{\text{in}})/(C_{\text{glc,in}} - C_{\text{glc,out}}) = D/q_{\text{glc}}$ where x is the biomass concentration (g dry biomass/l), C_{glc} is the appropriate glucose concentration (g glucose/l), D is the dilution rate (h^{-1}), and q_{glc} is the specific glucose consumption rate [g glucose/(g dry biomass$\times$h)].

3.5 Colony Biofilm Culturing

Microorganisms in biofilms often display phenotypes distinct from planktonic cells. The colony biofilm (CBF) method is a no-shear biofilm culturing technique which exhibits a relatively high degree of reproducibility [22]. Colony biofilms (*see* Fig. 5) serve as an excellent test system for engineered cross-feeding relationships due to the close physical proximity of the immobilized cells. Standardized methods for the CBF system and drop plate enumeration techniques have been described previously [22, 23].

1. Generate CBF inocula according to Subheading 3.3, **step 1**.

2. Prepare the CBF substratum by aseptically placing 1–3 polycarbonate membranes (*see* Subheading 2.3, **item 3**) onto M9 agar plates (standard 100×15 mm agar plates, with appropriate carbon source) using sterile forceps. Do not exceed three CBFs per plate.

3. Carefully pipette 100 µl of inoculum onto the center of each membrane. Allow agar/substratum to absorb liquid from

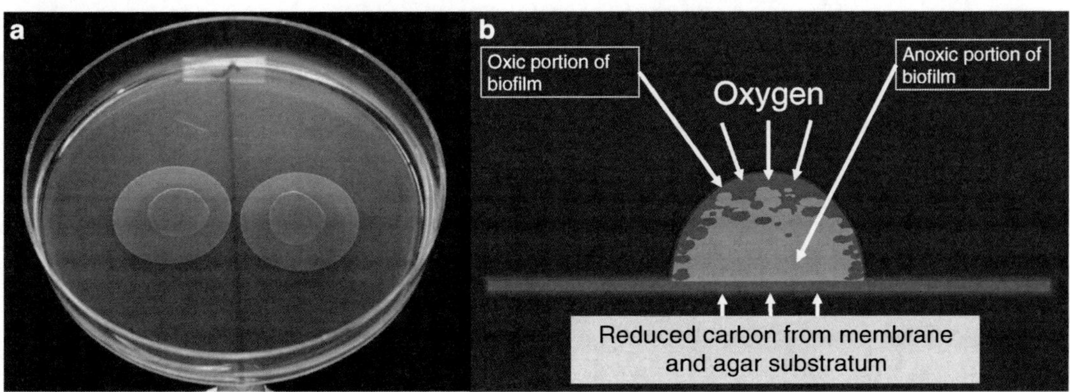

Fig. 5 (**a**) Two colony biofilms cultured on M9 minimal medium agar containing 2 g/l sodium acetate. (**b**) Schematic drawing of a colony biofilm illustrating the intrinsic nutrient gradients and inherent heterogeneity associated with aerobic culturing

inocula before transferring plates to 37 °C incubator (this helps eliminate the risk of inocula culture spreading from membrane). Once biofilms are firmly attached to membranes (1–5-h incubation), the plates may be incubated upside down to prevent accumulation of condensate.

4. Incubate CBFs for desired time period while aseptically transferring them to fresh M9 agar plates every 24 h with sterile forceps.

5. To begin processing CBF sample, aseptically transfer membrane with attached biofilm into a test tube with 5 ml sterile PBS. Vortex vigorously for 30 s to remove cells from membrane. Carefully remove membrane from cell suspension with sterile forceps.

6. Disaggregate cells with a tissue homogenizer at 16,000 rpm for 30 s. Use separate sterile homogenizing blades and/or treat with 70 % ethanol for 15 s followed by sterile PBS rinse between each sample.

7. The 5 ml, homogenized biofilm suspensions represent the zero dilution samples. Serially dilute samples 1:10 with sterile PBS (1 ml sample added to 9 ml PBS). Vigorously vortex each dilution sample prior to the next sequential transfer to ensure proper mixing. *E. coli* CBFs cultured for 0–168 h under this method are typically enumerated with the third to ninth dilution.

8. Carefully pipette ten individual 10 µl drops from each dilution onto LB agar (typically one agar plate can be sectioned into halves or thirds to accommodate drops from multiple dilutions) (*see* **Note 14**). Incubate *drop plates* at 37 °C for 6–20 h, and enumerate colony-forming units (CFU). Back-calculate through dilutions to determine the number of CFU per original biofilm.

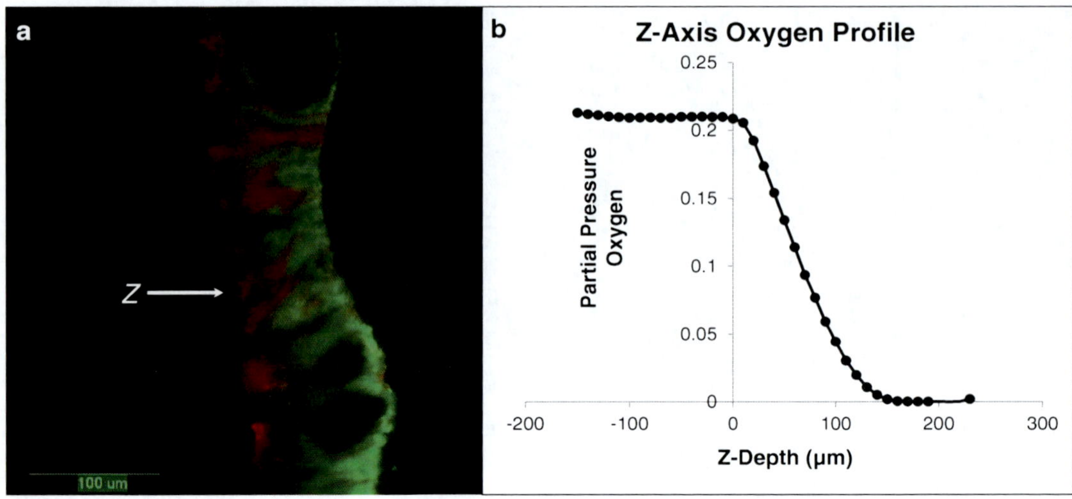

Fig. 6 (**a**) Epi-fluorescence micrograph of cryosectioned colony biofilm composed of two engineered *E. coli* mutants: obligate glucose consumer (*green*) and glucose-negative by-product oxidizer (*red*). (**b**) Oxygen microprofile measured in an analogous colony biofilm. The microprofile clearly shows portions of biofilm where aerobic and anaerobic microenvironments exist

3.6 Microscopy Preparation Methods

Microscopy can be a valuable tool for characterizing engineered microbial consortia grown as biofilm or planktonic communities. Here, we focus on microscopy preparation techniques useful for spatially characterizing CBF binary cultures expressing fluorescent reporter proteins (*see* **Note 15**). Figure 6a shows an example for a CBF composed from an engineered *E. coli* binary culture.

1. Culture CBFs according to the methods described in Subheading 3.5. Use strains containing different reporter plasmids (*see* Subheading 2.5, **item 3**). Culture biofilms on agar plates containing 100 μg/ml ampicillin or other antibiotics appropriate for the reporter plasmids.

2. Transfer each membrane containing a mature biofilm to a thin rectangular stainless steel coupon kept at room temperature with the biofilm side facing up.

3. Cover the top of biofilm with tissue-embedding medium (avoid air bubbles) (O.C.T., TissueTek), and immediately press the bottom of the stainless steel coupon onto dry ice. Do this quickly to minimize the spreading of the tissue-embedding medium prior to freezing.

4. Carefully remove membrane containing the frozen, embedded CBF from the steel coupon with clean forceps (avoid tearing membrane). Removal is sometimes aided by applying torque to the stainless steel coupon.

5. Place CBF upside down (membrane side up) on a flat dry ice surface. Carefully trim excess membrane from the edges of the

embedded area with razor blade. Clear trimmings before proceeding to the next step.

6. Cover the bottom side of the trimmed CBF with tissue-embedding medium (avoid air bubbles). Make sure that the entire bottom and perimeter of CBF are covered and allow sample to freeze. Sample may now be stored for short periods of time at −80 °C.

7. Transfer embedded CBF sample to a cryostat (*see* Subheading 2.5, **item 2**) and incubate at −20 °C for 30 min.

8. After incubation, cut embedded CBF sample in half with a clean razor blade. Attach one of the cut samples (cut edge facing out) to a cold (−20 °C) specimen disc with tissue-embedding medium.

9. Begin cryosectioning sample by cutting away the rough-cut edge until the entire cross section of the embedded CBF sample can be cut into single 5 μm thick slices.

10. Transfer selected 5 μm thick slices to the surface of Superfrost Plus microscope slide(s) (*see* Subheading 2.5, **item 2**). Samples are now ready to be viewed/imaged with the chosen microscopy technique (*see* **Note 5**).

3.7 Oxygen Microsensor Analysis

Biofilms are heterogeneous accretions of cells within extracellular polymeric substances (EPS) and possess gradients of chemical species (i.e., oxygen). Microbial community members often localize activity and physical location to specific regions along the oxygen gradient. Microsensor profiling is an excellent tool for spatially characterizing some microscale environmental conditions. Oxygen microsensors are well established, are a reasonably low-impact method for spatially quantifying concentration gradients [24, 25], and are especially powerful when combined with microscopy (methods highlighted in Subheading 3.6; Fig. 6b).

1. Culture CBFs according to the methods described in Subheading 3.5.

2. Pre-polarize oxygen microsensors for at least 5 h before use or until response to ambient air conditions becomes constant. This step removes the oxygen which may have accumulated in the sensors.

3. Perform a two-point calibration on sensor(s) immediately before use (*see* Subheading 2.6, **item 4**, and **Note 16**). Frequently check the calibration against known standards during the experiment to avoid error associated with signal drift.

4. Position the calibrated microsensor (mounted on motorized micromanipulator) just above the biofilm surface (typically 20–100 μm). Verify that the signal is appropriate for the partial pressure of oxygen in ambient air before profiling.

5. Begin profiling biofilm. Take measurements at regular time intervals appropriate to the sensor's specified response time. Set step size (distance traveled between measurements) for 1–2 times the tip diameter of the sensor for best-case spatial resolution.

6. Repeat profiles in multiple positions as necessary to account for biofilm heterogeneity. Process profiles in SensorTrace software or export data for custom processing (*see* **Notes 15** and **17**).

4   Notes

1. There are multiple software packages for performing EFMA which are updated continuously. Instructions and tutorials are typically included in the supporting literature. An example consortium model file with tutorial and descriptions of data processing procedures can be found in el-Mansi et al. [21]. This two-microbe model can be analyzed using the stand-alone pathway analysis software METATOOL and a spreadsheet program like MS Excel [13].

2. PCR primers should be designed to flank the gene to be deleted. Using DNA sequences from the coding regions of flanking genes often avoids nonunique DNA sequences found between genes. Internet-based resources like EcoGene (EcoGene.org) are helpful.

3. Here, the materials and methods for an inexpensive bioreactor design are presented. Figure 4 illustrates the custom-built reactor and basic components. Incubators, peristaltic pumps, and magnetic stir plates were used for temperature, flow, and agitation control, respectively. Similar reactors are commercially available from various suppliers such as DASGIP Technology (Eppendorf) or B. Braun.

4. Standards for HPLC analytes, ethanol and fumarate, are optional. Due to evaporation, ethanol is difficult to accurately measure in moderately to heavily sparged reactors. Due to its chemical structure, fumarate may cause complications with signal overlap when using HPLC-VWD chromatographs. Column incubation temperature can significantly affect the retention time of different analytes and can be modified to decrease signal overlap. More information on the Bio-Rad HPX-87H column (and other Aminex resin-based columns) can be found at the manufacturer's website (bio-rad.com).

5. The microscopy performed in the example study (Bernstein et al. 2012) was done with an epi-fluorescent microscope equipped with standard TRITC and FITC filters; however, in many cases scanning confocal laser microscopy may be a better

option for high-quality images especially for cryosectioned biofilms. Image analysis techniques may vary depending on the application. Our studies typically use Imaris (Bitplane) and Metamorph (Molecular Devices) image analysis software.

6. Once lysis has occurred, the P1 lysate should be processed within a couple of hours or else phage-resistant microbe variants will proliferate.

7. This incubation step allows time for antibiotic resistance gene expression. Typically the incubation times are 30 and 60 min for ampicillin and kanamycin, respectively.

8. Maintenance of the FT-A plasmid (*pFTA*) is temperature sensitive. Cultures containing it must be kept at or below 30 °C. Incubation at 37 °C will yield cultures which no longer replicate *pFTA* and lose their ampicillin resistance.

9. Successful gene deletions can typically be verified by kanamycin resistance screening (*see* Subheading 3.2, **step 18**) in combination with PCR. However, appropriate physiological tests (i.e., presence/absence of function associated with gene deletion) should also be performed when possible.

10. Transition times associated with lag, exponential, and stationary growth phases for engineered *E. coli* mutants may not be consistent among each other or with wild-type cultures. It is recommended that estimates for the maximum specific growth rates (and associated doubling times) be obtained through batch studies prior to chemostat growth. Simple shake flask experiments are useful for obtaining preliminary growth rate estimates. The authors typically set the dilution rates to $\leq$90 % of the measured maximum specific growth rate of the slowest growing community member to avoid chemostat *washout*.

11. High-to-moderate rates of air sparging (>0.5 volumes per min; VPM) may cause significant evaporation resulting in temperature changes within the reactor. A solution is to use an in-line humidifier incubated at the same temperature as the reactor. Higher rates of air sparging require higher humidifier volumes. Typically, 1 l of sterile DI water, incubated at 37 °C, is sufficient for up to 4 days of air sparging at or below 1 LPM.

12. For batch reactor operation, sample the minimum required volume for physiological measurements. Attempt to cumulatively sample less than 10 % of the culture volume over the course of the experiment. For chemostat operation, sample reactor regularly until OD_{600}, pH, and HPLC analyte measurements remain relatively constant. A practical definition of reactor chemostasis and metabolic steady state is when respective measurements stay within ±5 % of the steady-state mean value(s) for $\geq$3 residence times (residence time = 1/D).

13. Optical density (OD_{600}) is a measurement based primarily on light scattering which changes with cell density and cell geometry. Two important factors should be noted when taking optical density measurements and building cell dry weight standard curves: (a) OD_{600} values above 0.3 are less accurate and should be diluted (typically tenfold dilutions) to ensure a linear relationship with cell density and (b) cell dry weight standard curves often change with different physiological conditions and gene deletions. It is important to use cells taken from chemostats at chemostasis (steady state) and only apply standard curves to their respective growth condition and strain composition.

14. Bulk enumeration values are typically reported in CFU per biofilm. The drop plate enumeration technique may be modified to approximate individual populations from the biofilm or the reactor culture. This can be done with antibiotic selection markers or with strain-specific physiological conditions. An example would be enumerating two sets of drop plates using two strain-specific carbon sources such as glucose and acetate (specific to the example given in Bernstein et al. 2012). Relative abundances of community members in a biofilm or a liquid culture may also be determined via microscopy methods if the strains are expressing reporter genes (*see* Subheading 3.6).

15. Many additional biofilm techniques (including microscopy and reaction–diffusion analyses) have been developed at the Center for Biofilm Engineering, Montana State University, and can be found in *Biofilms: The Hypertextbook* (www.biofilmbook.com).

16. Clark-type oxygen microsensors respond linearly with the partial pressure of oxygen. There are multiple methods for calibration. Factors such as temperature, ionic strength, and solubility should be considered carefully before choosing a calibration medium. Avoid calibrating in liquid for obtaining concentration values from gas phase and vice versa. The authors recommend performing all calibrations in terms of partial pressure or % saturation which can be converted to concentration values when the appropriate medium physical properties are known.

17. Oxygen microprofiles in CBFs (or any immobilized biological matrix) can be interpreted with classical reaction and diffusion theory. Biofilm transport theory has been described in many scientific studies and remains a field of keen interest [26–28].

Acknowledgments

This work was funded by National Institute of Health grant (EB006532 and P20 RR024237) and the National Science Foundation-Integrative Graduate Education and Research Training (IGERT) Program (DGE 0654336) for support to

H.C.B. The authors would also like to acknowledge Alissa Bleem, Reed Taffs, James Folsom, Trevor Zuroff, and Betsey Pitts for their efforts associated with developing and iterating on the methods described here.

References

1. Kneitel JM, Chase JM (2004) Trade-offs in community ecology: linking spatial scales and species coexistence. Ecol Lett 7(1):69–80. doi:10.1046/j.1461-0248.2003.00551.x

2. McMahon KD, Martin HG, Hugenholtz P (2007) Integrating ecology into biotechnology. Curr Opin Biotechnol 18(3):287–292. doi:S0958-1669(07)00056-0 [pii]10.1016/j.copbio.2007.04.007

3. Bernstein HC, Paulson SD, Carlson RP (2012) Synthetic Escherichia coli consortia engineered for syntrophy demonstrate enhanced biomass productivity. J Biotechnol 157(1):159–166. doi:10.1016/j.jbiotec.2011.10.001

4. Brenner K, You L, Arnold FH (2008) Engineering microbial consortia: a new frontier in synthetic biology. Trends Biotechnol 26(9):483–489. doi:S0167-7799(08)00171-6 [pii]10.1016/j.tibtech.2008.05.004

5. Taffs R, Aston J, Brileya K, Jay Z, Klatt C, McGlynn S, Mallette N, Montross S, Gerlach R, Inskeep W, Ward D, Carlson R (2009) In silico approaches to study mass and energy flows in microbial consortia: a syntrophic case study. BMC Syst Biol 3(1):114

6. Wintermute EH, Silver PA (2010) Emergent cooperation in microbial metabolism. Mol Syst Biol 6:407. doi:msb201066 [pii]10.1038/msb.2010.66

7. Wintermute EH, Silver PA (2010) Dynamics in the mixed microbial concourse. Genes Dev 24(23):2603–2614. doi:24/23/2603[pii]10.1101/gad.1985210

8. Zuroff TR, Curtis WR (2012) Developing symbiotic consortia for lignocellulosic biofuel production. Appl Microbiol Biotechnol 93(4):1423–1435. doi:10.1007/s00253-011-3762-9

9. Wolfe AJ (2005) The acetate switch. Microbiol Mol Biol Rev 69(1):12. doi:10.1128/Mmbr.69.1.12-50.2005

10. Klamt S, Stelling J, Ginkel M, Gilles ED (2003) FluxAnalyzer: exploring structure, pathways, and flux distributions in metabolic networks on interactive flux maps. Biogeosciences 19(2):261–269. doi:10.1093/bioinformatics/19.2.261

11. Terzer M, Stelling J (2006) Accelerating the computation of elementary modes using pattern trees algorithms in bioinformatics. In: Bücher P, Moret B (eds) Lecture notes in computer science. Springer, Heidelberg, pp 333–343. doi:10.1007/11851561_31

12. Terzer M, Stelling J (2008) Large-scale computation of elementary flux modes with bit pattern trees. Biogeosciences 24(19):2229–2235. doi:10.1093/bioinformatics/btn401

13. Pfeiffer T, Sanchez-Valdenebro I, Nuno JC, Montero F, Schuster S (1999) METATOOL: for studying metabolic networks. Biogeosciences 15(3):251–257. doi:10.1093/bioinformatics/15.3.251

14. Carlson R, Srienc F (2004) Fundamental Escherichia coli biochemical pathways for biomass and energy production: Identification of reactions. Biotechnol Bioeng 85(1):1–19. doi:10.1002/bit.10812

15. Carlson RP (2007) Metabolic systems cost-benefit analysis for interpreting network structure and regulation. Biogeosciences 23(10):1258–1264. doi:10.1093/bioinformatics/btm082

16. Baba T, Ara T, Hasegawa M, Takai Y, Okumura Y, Baba M, Datsenko KA, Tomita M, Wanner BL, Mori H (2006) Construction of Escherichia coli K-12 in-frame, single-gene knockout mutants: the Keio collection. Mol Syst Biol 2: 2006.0008. http://www.nature.com/msb/journal/v2/n1/suppinfo/msb4100050_S1.html

17. Datsenko KA, Wanner BL (2000) One-step inactivation of chromosomal genes in Escherichia coli K-12 using PCR products. Proc Natl Acad Sci U S A 97(12):6640–6645. doi:10.1073/pnas.120163297

18. Pósfai G, Koob MD, Kirkpatrick HA, Blattner FR (1997) Versatile insertion plasmids for targeted genome manipulations in bacteria: isolation, deletion, and rescue of the pathogenicity island LEE of the Escherichia coli O157:H7 genome. J Bacteriol 179(13):4426–4428

19. Miller JH (1972) Experiments in molecular genetics. Cold Spring Harbor Laboratory, Cold Spring Harbor, NY

20. Schuster S, Dandekar T, Fell DA (1999) Detection of elementary flux modes in biochemical networks: a promising tool for pathway analysis and metabolic engineering. Trends Biotechnol 17(2):53–60. doi:10.1016/s0167-7799(98)01290-6

21. El-Mansi M, Stephanopoulos G, Carlson RP (2011) Flux control analysis and stoichiometric network modeling: basic principles and industrial applications. In: El-Mansi M, Bryce CFA, Demian AL, Allman AR (eds) Fermentation microbiology and biotechnology. CRC/Taylor and Francis Inc., Oxford, UK, pp 150–190

22. Hamilton M (2003) The biofilm laboratory step-by-step protocols for experimental design, analysis, and data interpretation. Cytergy Publishing, Bozeman, MT

23. Herigstad B, Hamilton M, Heersink J (2001) How to optimize the drop plate method for enumerating bacteria. J Microbiol Meth 44(2):121–129. doi:S016770120000 2414 [pii]

24. Revsbech NP (1989) An oxygen microsensor with a guard cathode. Limnol Oceanogr 34(2):474–478

25. Revsbech NP, Jorgensen BB (1986) Microelectrodes – their use in microbial ecology. Adv Microb Ecol 9:293–352

26. Beyenal H, Lewandowski Z, Harkin G (2004) Quantifying biofilm structure: facts and fiction. Biofouling 20(1):1–23. doi:10.1080/08927 01042000191628

27. Stewart PS (2003) Diffusion in biofilms. J Bacteriol 185(5):1485–1491

28. Xu KD, Stewart PS, Xia F, Huang CT, McFeters GA (1998) Spatial physiological heterogeneity in Pseudomonas aeruginosa biofilm is determined by oxygen availability. Appl Environ Microbiol 64(10):4035–4039

Chapter 5

An Observation Method for Autonomous Signaling-Mediated Synthetic Diversification in *Escherichia coli*

Ryoji Sekine, Shotaro Ayukawa, and Daisuke Kiga

Abstract

Phenotypic diversification of cells in development and regeneration is conceptually modeled by the motion of marbles rolling down valleys on the Waddington landscape, the main feature of which is bifurcations of the valleys. We have experimentally shown that this feature is sufficient to achieve phenotypic diversification by the construction of a synthetic phenotypic diversification system in *Escherichia coli*. Cells containing the synthetic phenotypic diversification system were diversified into two distinct cell states, high and low, through autonomous signaling-mediated bifurcation, when all cells were initialized to the low state. In this chapter, we illustrate the detailed experimental procedures involved in the initialization of cells and the observation of the phenotypic diversification.

Key words Phenotypic diversification, Intercellular signaling, Epigenetic landscape, Bifurcation, Synthetic biology

1 Introduction

Cellular phenotypic diversification is the dynamic increase in the number of phenotypes during developmental and regenerative processes in multicellular organisms. The concept of cellular phenotypic diversification is represented as marbles rolling down valleys on Waddington's epigenetic landscape [1]. This simple concept is used to interpret the dynamic phenotypic changes of various organisms in development or diversification [2, 3]. The main feature of the landscape is the bifurcations of the valleys, indicating the increase in available phenotype(s).

We have recently proposed a synthetic diversification system (Fig. 1a) in *Escherichia coli* to confirm that the main feature of the landscape is sufficient to experimentally demonstrate phenotypic diversification [4]. The synthetic diversification system consists of two plasmids: pHT_luxI1.5C as the diversity generator plasmid and pLuxR as the LuxR production plasmid. The diversity generator

Lianhong Sun and Wenying Shou (eds.), *Engineering and Analyzing Multicellular Systems: Methods and Protocols*,
Methods in Molecular Biology, vol. 1151, DOI 10.1007/978-1-4939-0554-6_5, © Springer Science+Business Media New York 2014

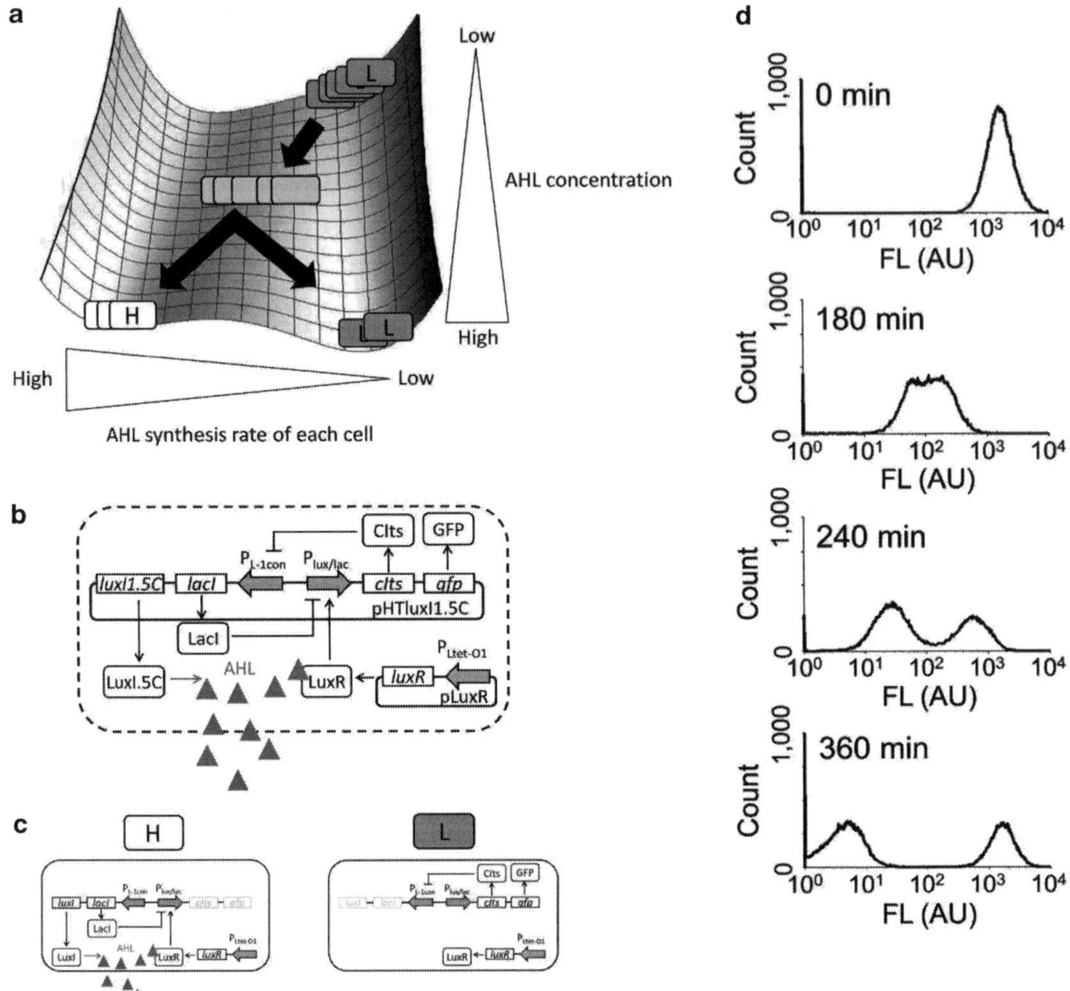

Fig. 1 The synthetic diversification system using the feature of Waddington landscape. (**a**) The horizontal axis represents AHL productivity. The vertical axis represents the AHL concentration. (**b**) The genetic network. Expression of both the LacI repressor protein and an AHL production enzyme LuxI1.5C is repressed by C1ts. Expression of both the C1ts repressor protein and green fluorescent protein GFPmut3 is repressed by LacI and activated by LuxR receptor protein, which itself is activated by AHL. The $P_{Ltet-O1}$ promoter in the pLuxR plasmid controls the expression of the LuxR. (**c**) The high-state cells produce LuxI1.5C and LacI, whereas the low-state cells produce C1ts and GFPmut3. (**d**) GFP fluorescence histograms of HT_luxI1.5C cells at measurement time points (0, 180, 240, and 360 min from the initialization) [4]. Populations with strong GFP fluorescence are the low-state cell populations, whereas populations with weak fluorescence are the high-state cell populations. FL indicates the GFP fluorescence intensity as collected by flow cytometry (Becton-Dickinson FACSCalibur)

plasmid has two key promoters: the $P_{L-1\ con}$ promoter, which is repressed by C1ts, and the $P_{lux/lac}$ promoter, which is repressed by LacI and activated by N-acyl homoserine lactone (AHL) (Fig. 1b). The $P_{L-1\ con}$ promoter regulates the expression of LacI and an AHL production enzyme LuxI1.5C, whereas the $P_{lux/lac}$ promoter regulates the expression of CIts and GFPmut3. Thus, cells containing the synthetic diversification system have the potential to become either of the two distinct cell states, namely, high and low states. We have defined the low state as the state where CIts and GFPmut3 are dominant and the high state as the state where LacI and LuxI1.5C are dominant (Fig. 1c). The accumulation of AHL produced by LuxI1.5C causes bifurcation of the diversification system: the high and low states are stable under a sufficient AHL concentration, whereas only the high state is stable under an insufficient AHL concentration. Therefore, if at the initial time all of the cells containing the synthetic diversification system are of the low state and the AHL concentration is low, the cells diversify into the two cell states through the bifurcation (Fig. 1d). To observe the synthetic diversification, the histograms of GFP fluorescence by flow cytometry can be measured.

In this chapter, we describe detailed protocols for the preparation of the low-state cells with low AHL concentration and the observation of the synthetic diversification.

2 Materials

2.1 Strain and Plasmids

1. Strain: *E. coli* strain JM 2.300 [λ-, lacI22, rpsL135 (StrR), thi-1] [5].

2. Plasmids: Diversity generator pHT_luxI1.5C and pLuxR for the constitutive LuxR protein production (Fig. 1b). The luxI1.5C gene on the pHT_luxI1.5C contains A117T and A188G mutations, which are a subset of mutations in previously reported luxI variant [6].

2.2 Medium and Reagents

1. LB liquid medium is sterilized by autoclaving and stored at room temperature.

2. 25 mg/mL stock solution of carbenicillin is sterilized using a 0.22-μm filter and stored at –20 °C.

3. 25 mg/mL stock solution of kanamycin is sterilized using a 0.22-μm filter and stored at –20 °C.

4. 1 mM AHL (3OC6HSL) is prepared in dimethyl sulfoxide (DMSO), and 105 μL aliquots are stored at –80 °C.

5. 100 mM isopropyl β-D-1-thiogalactopyranoside (IPTG) is prepared in Milli-Q water and sterilized using a 0.22-μm filter, and 2-mL aliquots are stored at –20 °C.

6. 10× stock solution of phosphate-buffered saline (PBS) is stored at room temperature.

3   Methods

3.1   Initialization of the Cells to the Low State

1. Inoculate a 15 % glycerol stock of cells containing the pHT_luxI1.5C and pLuxR plasmids into 3 mL of LB liquid medium with antibiotics (50 µg/mL carbenicillin and 30 µg/mL kanamycin), referred to as "basal medium," in a 14-mL polypropylene round-bottom test tube (*see* **Note 1**).

2. Incubate the cells at 37 °C with shaking at 180 rpm for approximately 12 h.

3. Dilute the 10× PBS stock solution with Milli-Q water to make a standard solution and sterilize using a 0.22-µm filter. Filtered PBS is stored at 4 °C.

4. Prepare 50 mL of the basal medium. Store at 32 °C.

5. Dilute the culture 100-fold into 3 mL of the basal medium.

6. Incubate the cells at 37 °C with shaking at 180 rpm until all samples reach a sufficient density ($OD_{590} \geq 0.35$) (*see* **Note 2**).

7. Adjust optical densities of all samples to 0.35 by dilution with the basal medium.

8. Wash a 1-mL portion of the culture with sufficient density by using 1 mL of fresh basal medium (*see* Subheading 3.3).

9. Inoculate the washed culture into 100 mL of fresh basal medium containing 1 µM AHL and 2 mM IPTG in an autoclaved 500-mL triangle flask with baffle.

10. Incubate the cells at 32 °C with shaking at 140 rpm for 240 min (*see* **Notes 3** and **4**).

11. Sample a 1-mL portion of the culture, and measure the GFP fluorescence (*see* Subheading 3.4).

3.2   Observation of the Diversification

1. Prepare 50 mL of the basal medium. Store at 32 °C.

2. Wash a 2-mL portion of the low-state initialized cell culture prepared in Subheading 3.1 three times with 1 mL of fresh basal medium to remove the IPTG and AHL from the culture (*see* **Note 5**).

3. Dilute the washed culture to the target density, referred to as the "initial cell density," into 100 mL of fresh basal medium in an autoclaved 500-mL triangle flask with baffle (*see* **Note 6**).

4. Incubate the culture at 32 °C with shaking at 140 rpm for 360 min.

5. Sample a portion of the culture at 180, 240, and 360 min (*see* **Note 7**), and measure the GFP fluorescences (*see* **Note 8**).

3.3   Culture Washing

1. Place a 1- or a 2-mL portion of the culture into a sterilized 2-mL tube.

2. Centrifuge the tube at $5,000 \times g$ for 1 min at 25 °C.

3. Remove the supernatant in the tube.

4. Resuspend the pellet in 1 mL of fresh basal medium warmed to 32 °C.

3.4 Fluorescence Measurements

1. Place a culture into an appropriate tube (*see* **Note 8**).

2. Centrifuge the tube at 9,000 × g for 1 min at 4 °C.

3. Remove the supernatant in the tube.

4. Resuspend the pellet in 1 mL of filtered PBS and transfer to a 5-mL polystyrene tube (BD Falcon) through a 40-μm cell strainer (BD Falcon).

5. Measure the GFP fluorescence of the prepared sample by flow cytometry (Becton-Dickinson FACSCalibur) using a 488 nm laser and a 515–545 nm emission filter.

---

4 Notes

1. The basal medium is prepared at three time points: the first just before Subheading 3.1, **step 1**; the second at Subheading 3.1, **step 4**; and the third at Subheading 3.2, **step 1**. The prepared medium is stored at 32 °C.

2. After approximately 120 min, OD_{590} will reach the sufficient value ($OD_{590} \cong 0.35$).

3. Due to the temperature sensitivity of C1ts on the diversity generator, the incubation temperature is set at 32 °C.

4. At this point, OD_{590} reaches between 0.30 and 0.40.

5. This procedure should be finished as soon as possible because time delay in this procedure results in fluctuations in the experiment.

6. The initial cell density for the diversification is approximately 3.0×10^5 (cells/mL). The volume of culture to dispense, V (mL), is calculated using the following equation:

$$V = \frac{V_m}{2.0 \times D} \times D_{init}$$

where V_m (100 mL) is the volume of the fresh basal medium, D (cells/mL) is the cell density of the initialized cell culture, and D_{init} (cells/mL) is the target cell density. The cell density of the initialized cell culture is estimated using the equation $D = OD_{590} \times 1.0 \times 10^9$ (cells/mL), where OD_{590} is the optical density of the cell culture.

7. To obtain a sufficient amount of cells for the measurement, 10, 5, and 1 mL of culture are sampled at 180, 240, and 360 min, respectively.

8. The proportion of the low-state cells in the population at 360 min will be 39 ± 1 %.

References

1. Waddington CH (1957) The strategy of the genes. George Allen & Unwin, London
2. Mohammad HP, Baylin SB (2010) Linking cell signaling and the epigenetic machinery. Nat Biotechnol 28:1033–1038
3. Balázsi G, van Oudenaarden A, Collins JJ (2011) Cellular decision making and biological noise: from microbes to mammals. Cell 144:910–925
4. Sekine R, Yamamura M, Ayukawa S et al (2011) Tunable synthetic phenotypic diversification on Waddington's landscape through autonomous signaling. Proc Natl Acad Sci USA 108: 17969–17973
5. Gardner TS, Cantor CR, Collins JJ (2000) Construction of a genetic toggle switch in *Escherichia coli*. Nature 403:339–342
6. Kambam PK, Sayut DJ, Niu Y et al (2008) Directed evolution of LuxI for enhanced OHHL production. Biotechnol Bioeng 101:263–272

Chapter 6

Integration-Free Reprogramming of Human Somatic Cells to Induced Pluripotent Stem Cells (iPSCs) Without Viral Vectors, Recombinant DNA, and Genetic Modification

Boon Chin Heng and Martin Fussenegger

Abstract

Stem cells are envisaged to be integral components of multicellular systems engineered for therapeutic applications. The reprogramming of somatic cells to induced pluripotent stem cells (iPSCs) via recombinant expression of a limited number of transcription factors, which was first achieved by Yamanaka and colleagues in 2007, heralded a major breakthrough in the stem cell field. Since then, there has been rapid progress in the field of iPSC generation, including the identification of various small molecules that can enhance reprogramming efficiency and reduce the number of different transcription factors required for reprogramming. Nevertheless, the major obstacles facing clinical applications of iPSCs are safety concerns associated with the use of viral vectors and recombinant DNA for expressing the appropriate transcription factors during reprogramming. In particular, permanent genetic modifications to newly reprogrammed iPSCs have to be avoided in order to meet stringent safety requirements for clinical therapy. These safety challenges can be overcome by new technology platforms that enable cellular reprogramming to iPSCs without the need to utilize either recombinant DNA or viral vectors. The use of recombinant cell-penetrating peptides and direct transfection of synthetic mRNA encoding appropriate transcription factors have both been shown to successfully reprogram somatic cells to iPSCs. It has also been shown more recently that the direct transfection of certain miRNA species can reprogram somatic cells to pluripotency without the need for any of the transcription factors commonly utilized for iPSC generation. This chapter describes protocols for iPSC generation with these new techniques, which would obviate the use of recombinant DNA and viral vectors in cellular reprogramming, thus avoiding permanent genetic modification to the reprogrammed cells.

Key words miRNA, Pluripotency, Recombinant, Reprogramming, Transcription factors, Transfection, Stem cells

---

1 Introduction

Stem cells are envisaged to be integral components in the design and engineering of multicellular systems for therapeutic applications due to their multi-lineage differentiation potential and self-renewal capacity. The advent of induced pluripotent stem cells (iPSCs) in 2007 heralded a major breakthrough in the field of

Lianhong Sun and Wenying Shou (eds.), *Engineering and Analyzing Multicellular Systems: Methods and Protocols,*
Methods in Molecular Biology, vol. 1151, DOI 10.1007/978-1-4939-0554-6_6, © Springer Science+Business Media New York 2014

regenerative medicine [1, 2]. By utilizing the recombinant expression of just four transcription factors (Klf4, Oct4, c-Myc, and Sox2), Yamanaka and colleagues were able to reprogram mature somatic cells to a pluripotent embryonic stem cell-like state [1]. This discovery not only surmounted major ethical problems associated with harvesting pluripotent stem cells from human embryos [3] but also opened up the possibility of generating immunocompatible patient-specific stem cells for transplantation/transfusion therapy [4].

Ever since Yamanaka et al.'s pioneering study, there has been rapid progress in the iPSC field [5, 6]. Apart from the original four transcription factors (Klf4, Oct4, c-Myc, and Sox2) that Yamanaka et al. utilized initially, another additional two transcription factors (Nanog and Lin28) have also been shown to be able to substitute some of the original four transcription factors in cellular reprogramming for iPSC generation [7–10]. Various small molecules have also been identified that can enhance reprogramming efficiency and reduce the number of different transcription factors required for reprogramming to iPSCs [11–13]. Lin et al. [12] reported that utilizing a combination of three small molecules (SB431542, PD032-5901, and thiazovivin) produced a 200-fold increase in iPSC reprogramming efficiency. Another study by Li et al. [13] showed that the expression of only just two transcription factors (Oct4 and Klf4) was sufficient to reprogram primary human keratinocytes to iPSCs in the presence of two small molecules (CHIR99021 and Parnate). To date, however, the use of small molecules alone has not yet been shown to be sufficient for reprogramming somatic cells to the pluripotent state.

Currently, the overwhelming majority of studies on iPSC reprogramming utilize recombinant DNA and viral vectors, which carry the attendant risk of permanent genetic modification to the reprogrammed cells. This poses a major barrier to clinical applications of iPSCs, as there are serious safety concerns regarding the transplantation/transfusion of genetically modified cells into the human body. Indeed, in a previous clinical trial, genetically modified cells became cancerous, which led to the unfortunate death of the patient [14]. Although strategies to transiently insert and remove transgenic elements in a precise and site-specific manner have been developed with the PiggyBac transposon system [15, 16], such an approach is tedious and labor intensive and requires rigorous screening to confirm the absence of genetic modification to the cellular genome.

New technology platforms have recently been developed to enable cellular reprogramming to iPSCs without the need to utilize either recombinant DNA or viral vectors. The use of recombinant cell-penetrating peptides [17, 18] has been shown to successfully reprogram somatic cells to iPSCs. However, this approach is hampered by extremely low efficiency, partly due to the short half-life of cell-penetrating peptides upon entry into the cell [17–19]. A more efficient technique would be to transfect synthetic mRNA encoding the appropriate transcription factors [20, 21] into the cell for reprogramming to iPSCs. A major limitation of this approach is that artificial entry of synthetic mRNA into cells may trigger the innate immunity pathway that could lead to apoptosis. Although inhibitors of the innate immunity pathway such as B18R protein can mitigate apoptosis [22], a high proportion of mammalian cells still die off upon repeated transfection with synthetic mRNA in the presence of B18R [23]. Another limitation is the relatively high toxicity of chemical agents such as lipofectamine, which are utilized for repeated multiple transfections of synthetic mRNA during the reprogramming process [20, 21]; this high toxicity is another major cause of low cell survivability with this technique [23]. More recently, the direct transfection of certain miRNA species alone has been shown to be able to reprogram somatic cells to pluripotency [24] without the need for any of the transcription factors that are commonly utilized for iPSC generation. The major advantage of miRNA-based reprogramming is that the innate immunity pathway leading to apoptosis is not activated upon artificial entry of miRNA into mammalian cells on account of their small size [23]. Additionally, because there is no need for the transfected miRNA to be translated into proteins, its effects are much faster. Also, some miRNA species utilized in reprogramming can target various epigenetic factors simultaneously, which leads to global demethylation in target cells [25–27].

This chapter describes protocols for iPSC generation using the abovementioned new technology platforms, which would obviate the use of recombinant DNA and viral vectors in cellular reprogramming. This would ensure that the newly reprogrammed cells are completely free of genetic modification, thereby partially fulfilling the stringent safety requirements for clinical therapy (*see* **Note 1**).

2 Materials

Unless otherwise stated, all chemical reagents (analytical grade) were purchased from Sigma-Aldrich Inc. (St. Louis, MO, USA), all plastic labware consumables were purchased from Becton-Dickinson Inc. (Franklin Lakes, NJ, USA), and all cell culture media, serum, and serum supplements were purchased from Invitrogen Inc. (Carlsbad, CA, USA).

2.1 Various Somatic Cell Types for Reprogramming

1. Primary human neonatal foreskin fibroblasts (*see* **Note 2**):
   - American Type Culture Collection (Manassas, VA, USA) Cat No.

  - CRL-1634, CRL-2076, CRL-2088, CRL-2091, CRL 2094, CRL-2097, CRL-2114, CRL-2127, CRL-2429, CRL-2450, CRL-2522, CRL-2703, CRL-7026, CRL-7065.
- Lonza Inc. (Basel, Switzerland) Cat No.: CC-2509.

2. Primary human adult dermal fibroblasts:
   - American Type Culture Collection (Manassas, VA, USA) Cat No.: PCS-201-012.
   - Lonza Inc. (Basel, Switzerland) Cat No.: CC-2511.

3. Primary human bone marrow-derived mesenchymal stem cells:
   - Lonza Inc. (Basel, Switzerland) Cat No. : PT-2501.

4. Primary human adipose-derived mesenchymal stem cells:
   - American Type Culture Collection (Manassas, VA, USA) Cat No.: PCS-500-011.
   - Lonza Inc. (Basel, Switzerland) Cat No.: PT-5006.

2.2 Culture and Expansion of Somatic Cells

1. Primary human fibroblasts (adult dermal and neonatal foreskin) culture medium: High-glucose Dulbecco's minimum essential medium supplemented with 10 % (v/v) fetal bovine serum and 1 % (v/v) penicillin–streptomycin antibiotic solution.

2. For primary human mesenchymal stem cells (bone marrow and adipose derived), the specially formulated Mesenchymal Stem Cell Growth Medium (MSCGM®) from Lonza Inc. (Basel, Switzerland) is utilized.

3. Porcine gelatin solution: 0.1 % (w/v) (autoclaved).

4. Trypsin/EDTA solution: 0.05 % (w/v).

5. Ca^{2+}-free phosphate-buffered saline.

6. T-75 flasks.

2.3 iPSC Generation with Cell-Penetrating Peptides

1. Recombinant cell-permeable transcription factors: Stemgent® Recombinant Human Oct4-11R, Stemgent® Recombinant Human Sox2-11R, Stemgent® Recombinant Human Klf4-11R, and Stemgent® Recombinant Human c-Myc-11R (Stemgent, Cambridge, MA, USA).

2. Geltrex® LDEV-Free human embryonic stem cell (hESC)-qualified Reduced Growth Factor Basement Membrane Matrix (Invitrogen, Carlsbad, CA, USA).

3. Mitotically inactivated human neonatal fibroblast feeder (NuFF) cells treated with either gamma irradiation or mitomycin C (Global Stem, Rockville, MD, USA).

4. hESC medium: Knockout DMEM/F12 (Invitrogen, Carlsbad, CA, USA) supplemented with 20 % (v/v) Knockout Serum

Replacement (Invitrogen, Carlsbad, CA, USA), 10 ng/ml of basic Fibroblast Growth Factor (bFGF) (Invitrogen, Carlsbad, CA, USA), nonessential amino acids, 0.1 mM 2 β-mercaptoethanol, and 2 mM L-glutamine.

5. Ca^{2+}-free phosphate-buffered saline.

6. Six-well culture plate.

2.4 iPSC Generation Through Transfection with Synthetic Modified mRNA

1. Synthetic modified mRNA: Stemgent® Human Oct4 mRNA, Stemgent® Human Klf4 mRNA, Stemgent® Human Sox2 mRNA, Stemgent® Human Lin 28 mRNA, Stemgent® Human c-Myc mRNA, and Stemgent® nGFP mRNA (Stemgent, Cambridge, MA, USA). Chemical modifications to these mRNA molecules include addition of a 5′-guanine cap, removal of 5′-triphosphates, substitution of uridine with pseudouridine, and substitution of cytidine with 5-methylcytidine [20].

2. Geltrex® LDEV-Free hESC-qualified Reduced Growth Factor Basement Membrane Matrix (Invitrogen, Carlsbad, CA, USA).

3. Mitotically inactivated human NuFF cells treated with gamma irradiation or mitomycin C (Global Stem, Rockville, MD, USA).

4. Pluriton® reprogramming medium (Stemgent, Cambridge, MA, USA).

5. B18R recombinant protein carrier-free (eBioscience, San Diego, CA, USA).

6. Rock inhibitor Y27632 (Stemgent, Cambridge, MA, USA).

7. Opti-MEM® reduced serum medium (Gibco-BRL, Long Island, NY, USA).

8. Lipofectamine RNAiMAX® (Invitrogen, Carlsbad, CA, USA).

9. Recombinant bFGF (Invitrogen, Carlsbad, CA, USA).

10. Ca^{2+}-free phosphate-buffered saline.

11. hESC medium: Knockout DMEM/F12 supplemented with 20 % (v/v) Knockout Serum Replacement, 10 ng/ml of bFGF (Invitrogen Cat No. PHG0026), nonessential amino acids, 0.1 mM 2-mercaptoethanol, and 1 mM L-glutamine.

12. Six-well culture plate.

13. T-75 flasks.

14. Nuclease-free water.

2.5 iPSC Generation Through Transfection with miRNA

1. The miRNAs utilized for reprogramming of somatic cells to iPSCs consist of seven different species, the sequences of which are shown in Table 1. These species are hsa-mir-200c (MIMAT0000617), hsa-mir-302a (MIMAT0000684),

Table 1
miRNA sequences used for reprogramming somatic cells to iPSCs

miRNA	miRBase accession number	Sequence
hsa-mir-200c	MIMAT0000617	5′-UAA UAC UGC CGG GUA AUG AUG GA-3′
hsa-mir-302a	MIMAT0000684	5′-UAA GUG CUU CCA UGU UUU GGU GA-3′
hsa-mir-302b	MIMAT0000715	5′-UAA GUG CUU CCA UGU UUU UAG UAG-3′
hsa-mir-302c	MIMAT0000717	5′-UAA GUG CUU CCA UGU UUC AGU GG-3′
hsa-mir-302d	MIMAT0000718	5′-UAA GUG CUU CCA UGU UUG AGU GU-3′
hsa-mir-369-3p	MIMAT0000721	5′-AAU AAU ACA UGG UUG AUC UUU-3′
hsa-mir-369-5p	MIMAT0001621	5′-AGA UCG ACC GUG UUA UAU UCG C-3′

hsa-mir-302b (MIMAT0000715), hsa-mir-302c (MIMAT00 00717), hsa-mir-302d (MIMAT0000718), hsa-mir-369-3p (MIMAT0000721), and hsa-mir-369-5p (MIMAT0001621). These miRNAs of relatively short sequences can be custom-synthesized from various companies such as Sigma-Aldrich Inc. (St. Louis, MO, USA) and Applied Biosystems Inc. (Foster City, CA, USA).

2. Geltrex® LDEV-Free hESC-qualified Reduced Growth Factor Basement Membrane Matrix (Invitrogen, Carlsbad, CA, USA).

3. Opti-MEM® reduced serum medium (Gibco-BRL, Long Island, NY, USA).

4. Lipofectamine RNAiMAX® (Invitrogen, Carlsbad, CA, USA).

5. 0.05 % Trypsin/EDTA solution.

6. Ca^{2+}-free phosphate-buffered saline.

7. hESC medium: Knockout DMEM/F12 supplemented with 20 % (v/v) Knockout Serum Replacement, 10 ng/ml of bFGF (Invitrogen, Carlsbad, CA, USA), nonessential amino acids, 0.1 mM 2-mercaptoethanol, and 1 mM L-glutamine.

8. Nuclease-free water.

2.6   Culture and Expansion of Newly Reprogrammed iPSCs

1. Geltrex® LDEV-Free hESC-qualified Reduced Growth Factor Basement Membrane Matrix (Invitrogen, Carlsbad, CA, USA).

2. Mitotically inactivated human NuFF cells treated with gamma irradiation or mitomycin C (Global Stem, Rockville, MD, USA).

3. Rock inhibitor Y27632 (Stemgent, Cambridge, MA, USA).

4. 1 mg/ml Dispase solution (Stemcell Technologies, Vancouver, Canada).

5. Ca^{2+}-free phosphate-buffered saline.

6. hESC medium: Knockout DMEM/F12 supplemented with 20 % (v/v) Knockout Serum Replacement, 10 ng/ml of bFGF (Invitrogen, Carlsbad, CA, USA), nonessential amino acids, 0.1 mM 2-mercaptoethanol, and 1 mM L-glutamine.

2.7   Confirming the Pluripotency of Newly Reprogrammed iPSCs

1. Antibodies for detection of pluripotency markers: Rabbit anti-OCT3/4 IgG (Santa Cruz Biotechnology, Santa Cruz, CA, USA), Mouse anti-Nanog IgG (Santa Cruz Biotechnology, Santa Cruz, CA, USA), Mouse anti-TRA-1-60 IgG (Millipore, Billerica, MA, USA), and Mouse anti-SSEA-4 IgG (Millipore, Billerica, MA, USA). Antibodies for live staining of putative iPSC colonies: StainAlive® DyLightTM 488 TRA-1-60 and StainAlive DyLight® 488 TRA-1-81 (Stemgent, Cambridge, CA, USA).

2. Secondary antibodies—Goat anti-Mouse IgG FITC-conjugated and Goat anti-Rabbit IgG Rhodamine-conjugated (Millipore, Billerica, MA, USA).

3. 4 % (w/v) paraformaldehyde solution.

4. 0.1 % (v/v) Triton X-100 solution.

5. Goat serum.

6. Ca^{2+}-free phosphate-buffered saline.

7. 0.05 % Trypsin/EDTA solution.

8. StemPro® Accutase® Cell Dissociation Reagent (Invitrogen, Carlsbad, CA, USA).

9. PureLink® RNA mini kit (Invitrogen, Carlsbad, CA, USA).

10. PureLink® Dnase set (Invitrogen, Carlsbad, CA, USA).

11. Nuclease-free water.

12. Superscript® III First-Strand Synthesis System (Invitrogen, Carlsbad, CA, USA).

13. Taqman® Fast Advanced Mastermix (Invitrogen, Carlsbad, CA, USA).

14. Taqman® Gene Ex Assays (Invitrogen, Carlsbad, CA, USA) for pluripotency markers: OCT4 (Assay ID: Hs01895061_u1), Sox2 (Assay ID: Hs01053049_s1), Nanog (Assay ID: Hs04260366_g1), LIN28 (Assay ID: Hs00702808_s1), and hTERT (Assay ID: Hs00972656_m1).

15. Ultra low-attachment 6-well plates (Corning, Lowell, MA, USA).

16. Differentiation medium: KO-DMEM supplemented with 20 % (v/v) FBS (Invitrogen, Carlsbad, CA, USA), 2 mM L-glutamine, 0.1 mM nonessential amino acid, 1 % (v/v) penicillin–streptomycin antibiotic solution, and 0.1 mM 2-β-mercaptoethanol.

17. 0.1 % (w/v) porcine–gelatin solution (autoclaved).

18. 1 mg/ml Dispase solution (Stemcell Technologies, Vancouver, Canada).

19. 0.05 % (w/v) Trypsin/EDTA solution.

20. Antibodies for detection of differentiation markers of all three embryonic germ layers (endoderm, mesoderm, and ectoderm): mouse anti-α-fetoprotein, mouse anti-smooth muscle α-actin, and mouse anti-β3-tubulin.

21. Bovine serum albumin (BSA).

22. Vectashield® (Vector Labs, Burlingame, CA, USA).

3 Methods

Unless otherwise stated, all procedures were carried out at room temperature under sterile conditions within a laminar flow cabinet. All cell cultures and pre-coating of labware with gelatin or Geltrex® were carried out within a 5 % CO_2 incubator with low O_2 tension (3–5 % O_2) at 37 °C. It is necessary to use RNAase-free water, pipette tips, and microcentrifuge tubes for all experiments.

3.1 Culture and Expansion of Somatic Cells

1. Pre-coat T-75 flasks overnight with 0.1 % porcine–gelatin.

2. Seed 1.0–2.0×10^6 cells per T75 flask, constituted in the appropriate culture media according to the cell type.

3. Culture for 7–10 days within a 5 % CO_2 incubator at 37 °C until confluence is attained.

4. Dissociate confluent cell monolayer with 0.05 % trypsin–EDTA solution, wash in PBS, and reconstitute in fresh culture medium. Subculture into new T-75 flasks at a split ratio of 1:4–1:5.

3.2 iPSC Generation with Cell-Penetrating Peptides

1. Pre-coat 6-well plates with Geltrex® for 2 h, at the appropriate dilution according to the manufacturer's instructions.

2. Seed 5.0×10^4 of somatic cells per well of Geltrex®-coated 6-well plate.

3. Prepare 1 ml of reprogramming medium per well of 6-well plate. This involves reconstituting all four recombinant cell-permeable transcription factors (human Oct4-11R, human

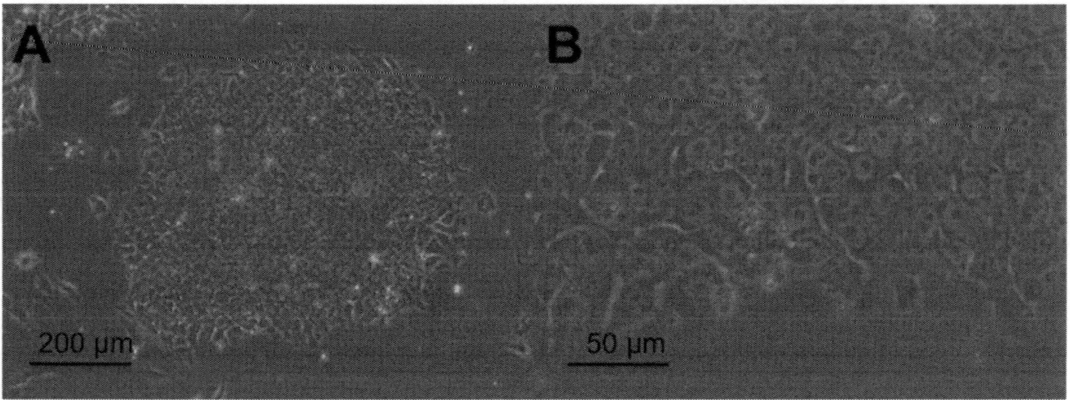

Fig. 1 (**a**) Typical colony of reprogrammed iPSCs observed under low-power magnification. (**b**) Typical colony of reprogrammed iPSCs observed under high-power magnification. Note the enlarged nuclei, reduced cytoplasm, and compact arrangement of cells within the colony

Sox2-11R, human Klf4-11R, and human c-Myc-11R) in hESC medium at a final concentration of 8 μg/ml for each protein [18].

4. Incubate the somatic cells overnight with 1 ml of reprogramming medium per well of 6-well plate for a total duration of 16 h.

5. The following day, replace the reprogramming medium with hESC medium.

6. Culture for a further 6 days with hESC medium, with a fresh medium change every day.

7. Repeat **steps 3–5** for a total of 6–8 cycles (*see* **Note 3**).

8. Individually pick up colonies with iPSC-like morphology (Fig. 1). Replate these onto NuFF feeders with hESC medium.

3.3 iPSC Generation Through Transfection with Synthetic Modified mRNA

1. Prior to starting the reprogramming experiment, NuFF-conditioned Pluriton® medium had to be prepared first. 4.0×10^6 NuFF cells (constituted in primary fibroblast culture medium) were seeded overnight into a T75 flask. The following day, the primary fibroblast culture medium was removed, the cells washed once with PBS, 25 ml of Pluriton® medium supplemented with 4 ng/ml of bFGF, and 1 % (v/v) of penicillin–streptomycin solution was then added into the T-75 flask. After 24-h incubation, the conditioned Pluriton® medium was collected, sterile-filtered, and frozen in appropriate aliquots at −20 °C, and fresh Pluriton® medium was added into the T-75 flask for another round of conditioning. This cycle was repeated up to six times.

2. A master mRNA reprogramming cocktail should also be prepared prior to starting the reprogramming experiment.

This should have a molar stoichiometry of 3:1:1:1:1:1 for the Oct4, Sox2, Klf4, c-Myc, Lin28, and nGFP mRNAs, respectively. This would correspond to 38.5 μg/ml Oct4 mRNA, 11.9 μg/ml Sox2 mRNA, 15.6 μg/ml Klf4 mRNA, 14.8 μg/ml c-Myc mRNA, 8.3 μg/ml Lin28 mRNA, and 11.1 μg/ml nGFP mRNA. Note that each mRNA factor (Stemgent, Cambridge, MA, USA) is at a concentration of 100 ng/μl. Once the cocktail is prepared, 50 μl aliquots should be stored at −80 °C. Each 50 μl aliquot is sufficient for reprogramming four wells of a 6-well plate.

3. Pre-coat 6-well plates with 0.1 % porcine–gelatin overnight.

4. Seed 2.5×10^5 NuFF cells (constituted in primary fibroblast culture medium) per well of the 6-well plate that was pre-coated with gelatin.

5. The following day, remove the primary fibroblast culture medium, wash once in PBS, and seed the target somatic cells (constituted in the appropriate culture medium) together with the adherent NuFF cells within the 6-well plate. The optimal seeding density for reprogramming with synthetic mRNA varies with different target somatic cell types and typically ranges from 5×10^3 to 2.5×10^4 cells per well of a 6-well plate. Therefore, a range of different seeding densities should be tried for the first reprogramming experiment.

6. The next day, the culture medium is replaced with 2 ml of Pluriton® medium supplemented with 200 ng/ml of B18R protein (*see* **Note 4**) per well of the 6-well plate. It is essential that the unsupplemented Pluriton® medium is first equilibrated for 2 h within the 5 % CO_2 incubator at 37 °C, to attain the appropriate pH and oxygen tension, prior to starting the reprogramming experiment. The B18R protein and Pluriton® medium supplement are added immediately before exposing the cells to the Pluriton® reprogramming medium.

7. The cells are ready to transfect after 2-h incubation with Pluriton® medium supplemented with 200 ng/ml of B18R protein. This is considered the day-0 time point.

8. The mRNA transfection complex is then prepared. This involves mixing 50 μl of the thawed master mRNA reprogramming cocktail with 200 μl of Opti-MEM in one tube and mixing 25 μl of RNAimax (*see* **Note 4**) with 225 μl of Opti-MEM medium in a second tube. The contents of the second tube are then added to the first tube, mixed through gentle pipetting, and left to stand for 15 min, so as to yield the mRNA transfection complex.

9. To each well, 120 μl of the mRNA transfection complex is added in a dropwise fashion. It is necessary to ensure uniform distribution of the mRNA transfection complex throughout

the entire well. This can be achieved by rocking the 6-well plate sideways and then backward and forward several times upon placement within the incubator.

10. The cells are then incubated with the mRNA transfection complex for a duration of 4 h.

11. After 4-h incubation, the medium in each well (containing the transfection complex) is removed and replaced with 2 ml of fresh equilibrated Pluriton® medium supplemented with 200 ng/ml of B18R protein. As before, it is essential that the unsupplemented Pluriton® medium is first equilibrated for 2 h within the 5 % CO_2 incubator at 37 °C to attain the appropriate pH and oxygen tension. The B18R protein and Pluriton® medium supplement are added just immediately before exposing the cells to the Pluriton® reprogramming medium.

12. Repeat **steps 8–10** for a further 5 days.

13. From day 6 to 20, **steps 8–10** are repeated again, except that the non-conditioned Pluriton® medium is replaced with the conditioned Pluriton® medium.

14. Individually pick up colonies with iPSC-like morphology (Fig. 1). Replate these onto NuFF feeders with hESC medium.

3.4 iPSC Generation Through Transfection with miRNA

1. A master miRNA reprogramming cocktail should also be prepared prior to starting the reprogramming experiment. A mixture of the seven different species of miRNA utilized for reprogramming (Table 1, hsa-mir-200c, hsa-mir-302a, hsa-mir-302b, hsa-mir-302c, hsa-mir-302d, hsa-mir-369-3p, and hsa-mir-369-5p) is dissolved in RNAse-free water to yield a molarity of 10 μM for each of the miRNA species. Once the cocktail is prepared, 50 μl aliquots should be stored at –80 °C. Each 50 μl aliquot is sufficient for reprogramming four wells of a 6-well plate.

2. Pre-coat 6-well plates with Geltrex® for 2 h at the appropriate dilution according to the manufacturer's instructions.

3. Seed 5.0×10^4 of target somatic cells (constituted in the appropriate culture media) per well of Geltrex®-coated 6-well plate.

4. The next day, the culture medium is replaced with 2 ml of Opti-MEM medium per well of the 6-well plate.

5. The miRNA transfection complex is then prepared. This involves mixing 50 μl of the thawed master miRNA reprogramming cocktail with 200 μl of Opti-MEM in one tube and mixing 25 μl of RNAimax (*see* **Note 5**) with 225 μl of Opti-MEM medium in a second tube. The contents of the second tube are then added to the first tube, mixed through gentle pipetting, and left to stand for 15 min, so as to yield the miRNA transfection complex.

6. 120 µl of the miRNA transfection complex is added to each well in a dropwise fashion. It is necessary to ensure uniform distribution of the miRNA transfection complex throughout the entire well. This can be achieved by rocking the 6-well plate sideways and then backward and forward several times upon placement within the incubator. This is considered the day-0 time point.

7. The cells are then incubated with the miRNA transfection complex for a duration of 4 h.

8. After 4-h incubation, the medium in each well (containing the transfection complex) is removed and replaced with 2 ml of fresh Opti-MEM medium and then cultured for a further 44 h (with replacement of fresh Opti-MEM medium after 24 h).

9. After 44 h of culture following transfection, **steps 4–8** are repeated for a further three cycles up to day 8 [24].

10. On day 6, pre-coat 6-well plates with 0.1 % porcine–gelatin overnight.

11. On day 7, seed 2.5×10^5 NuFF cells (constituted in primary fibroblast culture medium) per well of the 6-well plate that was pre-coated with gelatin.

12. On day 8, remove the primary fibroblast culture medium, wash once in PBS, and place 1 ml of hESC medium per well of 6-well plate seeded with NuFF cells.

13. On day 8, trypsinize the somatic cells transfected with miRNA, reconstitute in fresh hESC medium (1 ml per well of 6-well plate), and seed onto 6-well plate seeded with NuFF cells (one well per well).

14. Culture for a further 17–22 days, with daily replacement of fresh hESC medium (2 ml per well of 6-well plate).

15. Individually pick up colonies with iPSC-like morphology (Fig. 1). Replate these onto NuFF feeders with hESC medium.

3.5 Culture and Expansion of Newly Reprogrammed iPSCs

1. 2 days before passage and subculture of newly reprogrammed iPSCs, pre-coat 6-well plates with 0.1 % porcine–gelatin overnight.

2. The next day, seed 2.5×10^5 NuFF cells (constituted in primary fibroblast culture medium) per well of 6-well plate that was pre-coated with gelatin.

3. On the day of passage and subculture, remove the primary fibroblast culture medium, wash once in PBS, and place 2 ml of hESC medium per well of 6-well plate seeded with NuFF cells.

4. During the initial phase of culture and expansion of newly programmed iPSCs, when there are relatively few colonies, it is

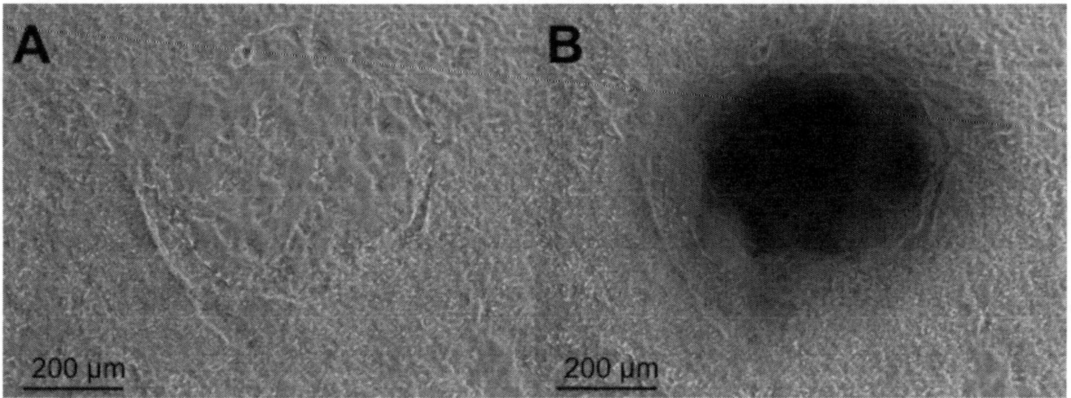

Fig. 2 (**a**) Area of differentiated cells within iPSC colony. (**b**) This can be marked from underneath the culture plate with a felt pen or marker under the view of an inverted microscope

preferable to pick up individual morphologically good-looking colonies of undifferentiated cells (Fig. 1) and seed them onto freshly plated NuFF feeder layers.

5. When there is sufficient quantity of cells (the well is confluent or almost confluent), enzymatic (dispase) dissociation of colonies into cellular clumps can be utilized for serial passage and subculture (*see* **Note 6**).

6. Prior to enzymatic dissociation, areas of differentiated cells (Fig. 2a) within individual wells should be removed by scraping with a pipette tip, under the view of a low-power stereomicroscope under sterile conditions within a laminar flow cabinet. Alternatively, if a stereomicroscope is unavailable, the 6-well plate can be viewed under a high-power inverted phase-contrast microscope, and the areas of differentiated cells within each individual well can then be marked out from underneath the culture plate with a felt pen or marker (Fig. 2b). The marked areas of differentiated cells can then be scraped off with a pipette tip under sterile conditions within a laminar flow cabinet.

7. After scraping off the differentiated cells, the hESC medium is removed from the well, and 0.5 ml of 1.0 mg/ml dispase solution is then added, followed by incubation for 5–10 min. The edges of individual colonies should appear to peel away (Fig. 3).

8. Remove the dispase solution from the well, and gently wash with 2 ml of PBS.

9. After removal of PBS, add in 2 ml of fresh hESC medium, and gently dislodge the colonies with gentle pipetting. When necessary, scraping with a pipette tip can also be used to dissociate

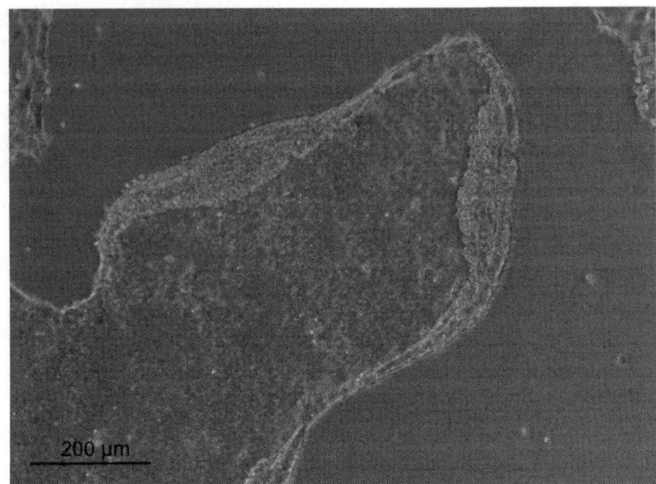

Fig. 3 Edges of iPSC colonies peel off after treatment with 1 mg/ml of dispase for 10–15 min

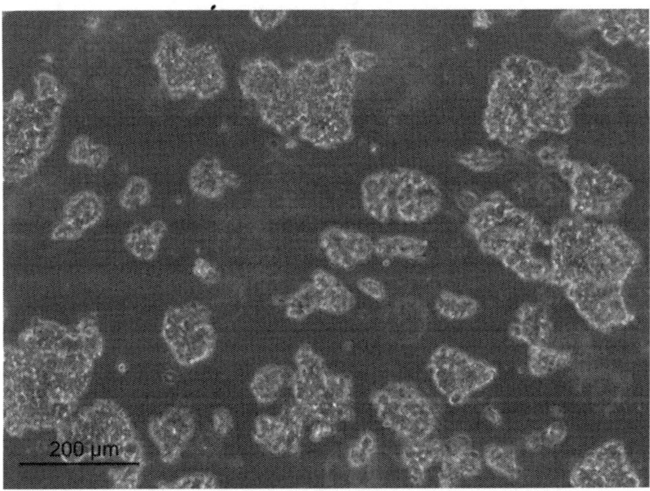

Fig. 4 Relatively large-sized iPSC clumps generated by 10–15 min of 1 mg/ml dispase treatment followed by gentle pipetting

colonies that are not dislodged through pipetting. Break up the dislodged colonies further into relatively large-sized cellular clumps (Fig. 4) and seed onto the freshly plated NuFF feeder layers.

10. Culture for 7–10 days with daily change of fresh hESC medium (2 ml per well), until confluence or near confluence is attained, prior to the next serial passage.

3.6 Confirming the Pluripotency of Newly Reprogrammed iPSCs

A variety of different techniques can be used to confirm the pluripotency of the newly reprogrammed iPSCs. These include live staining, immunocytochemistry, flow cytometry analysis, RT-PCR, and in vitro differentiation assays.

3.6.1 Live Staining of Newly Reprogrammed iPSC Colonies

1. Live staining can be utilized to identify and pick out newly reprogrammed iPSC colonies during the very first subculture and passage.

2. Dilute either the StainAlive® DyLightTM 488 TRA-1-60 antibody or StainAlive DyLight® 488 TRA-1-81 antibody in hESC medium to a final concentration of 5 µg/ml.

3. Remove culture medium from the well that is targeted for live staining. Add 1 ml of diluted antibody in hESC medium and incubate for 30 min within incubator.

4. Remove diluted antibody solution, wash with 2 ml of PBS, and then add 2 ml of hESC medium.

5. View under fluorescence microscopy (inverted microscope) with the appropriate excitation/emission wavelengths of 488 nm/520 nm, respectively. Identify fluorescently stained colonies.

6. Switch to bright-field microscopy, and mark out fluorescently stained colonies from underneath the culture plate with a felt pen or marker. The marked colonies can then be scraped off with a pipette tip under sterile conditions within a laminar flow cabinet and then reseeded on fresh NuFF feeder layers (Fig. 2).

3.6.2 Immunocytochemical Staining of Fixed Cells

1. Remove culture medium from the well that is targeted for immunocytochemical staining. Wash once with 2 ml of PBS, followed by addition of 0.5 ml of 4 % (w/v) paraformaldehyde solution. Fix cells overnight at 4 °C.

2. The following day, remove the 4 % (w/v) formaldehyde solution, and then wash once in 2 ml of PBS.

3. Remove PBS, and permeabilize cells by adding 0.5 ml of 0.1 % (v/v) Triton X-100 solution. Incubate at room temperature for 10 min. Then wash with 2 ml of PBS.

4. After removing PBS, add 0.5 ml of blocking solution consisting of 10 % (v/v) goat serum in PBS. Incubate at 37 °C for 1 h.

5. Remove blocking solution, wash with 2 ml of PBS, and then add 0.5 ml of diluted primary antibody solution against OCT 3/4, TRA-1-60, Nanog, or SSEA-4. Utilize PBS to dilute antibodies. Follow the manufacturer's instructions for the appropriate dilution of the primary antibodies.

6. Incubate fixed cells with the diluted primary antibodies for 1 h at 37 °C.

7. Remove diluted primary antibody solution, wash with 2 ml of PBS, and then add 0.5 ml of blocking solution consisting of 10 % (v/v) goat serum in PBS. Incubate at 37 °C for a further 1 h.

8. Remove blocking solution, wash with 2 ml of PBS, and then add 0.5 ml of appropriate secondary antibodies (fluorescently labeled) diluted in PBS. Again, follow the manufacturer's instructions for the appropriate dilution of the secondary antibodies.

9. Incubate fixed cells with the diluted secondary antibodies for 1 h at 37 °C.

10. Remove secondary antibody solution, wash once with 2 ml of PBS, and then add 0.5 ml of PBS into the well.

11. View and capture images under fluorescence microscopy with the appropriate excitation/emission wavelengths for the specific fluorophore conjugated to the secondary antibody.

3.6.3 Flow Cytometry Analysis of Pluripotency Markers

1. Remove culture medium from well, wash once with 2 ml of PBS, followed by addition of 1 ml of StemPro® Accutase® cell dissociation reagent, and incubate at 37 °C for 10–15 min within the incubator.

2. Dissociate into a single-cell suspension through gentle pipetting, and place within 15 ml tube.

3. Centrifuge down the cells at $1,000 \times g$ for 5 min, remove accustase solution, and resuspend in 0.5 ml of PBS.

4. Fix the cells through addition of 10 ml of 4 % (v/v) paraformaldehyde solution. Incubate overnight at 4 °C.

5. Centrifuge down the cells at $1,000 \times g$ for 5 min, remove the paraformaldehyde solution, and permeabilize cells by resuspending the fixed cells in 2 ml of 0.1 % (v/v) Triton X-100 solution. Incubate at room temperature for 10 min. Then centrifuge down the cells at $1,000 \times g$ for 5 min and resuspend in 2 ml of blocking solution consisting of 10 % (v/v) goat serum in PBS. Incubate at 37 °C for 1 h.

6. Centrifuge down the cells at $1,000 \times g$ for 5 min, remove blocking solution, and wash with 2 ml of PBS. Then resuspend cells in 2 ml of diluted primary antibody solution against OCT 3/4, TRA-1-60, Nanog, or SSEA-4. Utilize PBS to dilute antibodies. Follow the manufacturer's instructions for the appropriate dilution of the primary antibodies.

7. Incubate the fixed cells with the diluted primary antibodies for 1 h at 37 °C.

8. Centrifuge down the cells at $1,000 \times g$ for 5 min, remove diluted primary antibody solution, wash with 2 ml of PBS, and then resuspend cells in 2 ml of appropriate secondary antibodies

(fluorescently labeled) diluted in PBS. Again, follow the manufacturer's instructions for the appropriate dilution of the secondary antibodies.

9. Incubate fixed cells with the diluted secondary antibodies for 1 h at 37 °C.

10. Centrifuge down the cells at $1,000 \times g$ for 5 min, remove the secondary antibody solution, and resuspend in 1 ml of PBS.

11. The cells can now be analyzed with flow cytometry. Gates were typically set at the point of intersection between the negative and the positive stains, after which the percentage of cells from the negative control within the gate was subtracted from the positive.

3.6.4 RT-PCR Analysis for Expression of Pluripotency Markers

1. RNA from the putative iPSCs was harvested using the PureLink® RNA mini kit with DNase treatment, according to the manufacturer's instructions.

2. cDNA was synthesized from the extracted RNA, using the Superscript® III First-Strand Synthesis System, according to the manufacturer's instructions.

3. RT-PCR analysis can be performed with a programmable thermocycler apparatus, utilizing Taqman® Fast Advanced Mastermix, together with appropriate Taqman® Gene Ex Assays for specific pluripotency markers: OCT4 (assay ID: Hs01895061_u1), SOX2 (assay ID: Hs01053049_s1), Nanog (assay ID: Hs04260366_g1), LIN28 (assay ID: Hs00702808_s1), and hTERT (assay ID: Hs00972656_m1). The following amplification parameters can be utilized for the RT-PCR: 2 min at 50 °C, 20 s at 95 °C, and 40 cycles of 1 s at 95 °C, followed by 1 min at 60 °C. The relative cycle threshold (Ct) was determined and normalized against the endogenous GAPDH gene. The fold change of each gene was compared against the appropriate negative and positive control cell lines (i.e., parent somatic cell line and established hESC line, respectively).

3.6.5 In Vitro Differentiation Assay

1. The putative iPSCs were dissociated into relatively large-sized clumps (Fig. 4) by dispase and grown as embryoid bodies (EBs) on ultralow attachment 6-well plates in differentiation medium. Fresh differentiation medium was replaced every 2 days of culture.

2. Following 7 days of differentiation, the EBs were dissociated through trypsinization and subsequently replated onto gelatin-coated 6-well plates, upon which they were cultured for a further 14 days. Fresh differentiation medium was replaced every 2 days of culture.

3. Immunostaining was then carried out to identify cells from the three primary germ layers.

4. The differentiated iPSCs were fixed with 4 % (v/v) paraformaldehyde for 15 min and blocked for 2 h in PBS buffer containing 0.1 % Triton X-100, 10 % goat serum, and 1 % BSA.

5. The fixed cells were then incubated for 2 h at room temperature with the primary antibodies (anti-α-fetoprotein, anti-smooth muscle α-actin, anti-β3-tubulin) diluted in 1 % BSA/PBS at appropriate concentrations, according to the manufacturer's instructions.

6. The cells were then washed in 1 % BSA/PBS and incubated in the dark with the appropriate fluorophore-conjugated secondary antibodies for 2 h at room temperature. The secondary antibodies were diluted in 1 % BSA/PBS at appropriate concentrations, according to the manufacturer's instructions. Following another wash with 1 % BSA/PBS, a fluorescent mounting medium with DAPI (Vectashield®) was added to cover the cells and incubated for 1 h before observing the images under fluorescence microscopy.

4 Notes

1. Although the protocols described here enable the generation of iPSCs without the use of recombinant DNA and viral vectors and are therefore free of genetic modification, this is still insufficient to meet the rigorous safety standards for clinical applications. For clinical therapy, it is imperative that the cells are reprogrammed and cultured under xeno-free and feeder-free cGMP conditions.

2. Generally, human neonatal skin fibroblasts are easier to reprogram into iPSCs compared to somatic cells from more matured adult tissues. Hence, it would be preferable to include these cells as a parallel positive control, when attempting to reprogram somatic cells from more matured adult tissues into iPSCs.

3. The efficiency of generating iPSCs with cell-penetrating peptides is generally much lower than conventional methods that utilize recombinant DNA in the form of plasmids or viral vectors. This could be due to the relatively short half-life of the cell-penetrating peptides upon entry into the cell (*see* refs. 17–19).

4. The major challenge faced in generating iPSCs through transfection with synthetic modified mRNA is the activation of the interferon-based innate immunity pathway upon entry of mRNA into the cells. Although this can be partially mitigated by exposing cells to the interferon inhibitor B18R, a

high proportion of cells still die off due to apoptosis (*see* ref. 23). Another factor contributing to low cell survivability is the toxicity of lipofectamine utilized for mRNA transfection (*see* ref. 23).

5. Although miRNA transfection does not activate the innate immunity pathway within mammalian cells that lead to apoptosis, there is still a problem with toxicity of the lipofectamine utilized for transfection (*see* ref. 23).

6. Collagenase type IV 1 mg/ml can be used in place of dispase for serial passage and subculture of the reprogrammed iPSCs.

References

1. Takahashi K, Tanabe K, Ohnuki M, Narita M, Ichisaka T, Tomoda K, Yamanaka S (2007) Induction of pluripotent stem cells from adult human fibroblasts by defined factors. Cell 131:861–872

2. Yu J, Vodyanik MA, Smuga-Otto K, Antosiewicz-Bourget J, Frane JL, Tian S, Nie J, Jonsdottir GA, Ruotti V, Stewart R, Slukvin II, Thomson JA (2007) Induced pluripotent stem cell lines derived from human somatic cells. Science 318:1917–1920

3. Zacharias DG, Nelson TJ, Mueller PS, Hook CC (2011) The science and ethics of induced pluripotency: what will become of embryonic stem cells? Mayo Clin Proc 86:634–640

4. Fairchild PJ (2009) Transplantation tolerance in an age of induced pluripotency. Curr Opin Organ Transplant 14:321–325

5. Mostoslavsky G (2012) Concise review: the magic act of generating induced pluripotent stem cells: many rabbits in the hat. Stem Cells 30:28–32

6. Sidhu KS (2011) New approaches for the generation of induced pluripotent stem cells. Expert Opin Biol Ther 11:569–579

7. Zhao HX, Li Y, Jin HF, Xie L, Liu C, Jiang F, Luo YN, Yin GW, Li Y, Wang J, Li LS, Yao YQ, Wang XH (2010) Rapid and efficient reprogramming of human amnion-derived cells into pluripotency by three factors OCT4/SOX2/NANOG. Differentiation 80:123–129

8. Yan X, Qin H, Qu C, Tuan RS, Shi S, Huang GT (2010) iPS cells reprogrammed from human mesenchymal-like stem/progenitor cells of dental tissue origin. Stem Cells Dev 19:469–480

9. Li Y, Zhao H, Lan F, Lee A, Chen L, Lin C, Yao Y, Li L (2010) Generation of human-induced pluripotent stem cells from gut mesentery-derived cells by ectopic expression of OCT4/SOX2/NANOG. Cell Reprogram 12:237–247

10. Si-Tayeb K, Noto FK, Sepac A, Sedlic F, Bosnjak ZJ, Lough JW, Duncan SA (2010) Generation of human induced pluripotent stem cells by simple transient transfection of plasmid DNA encoding reprogramming factors. BMC Dev Biol 10:81

11. Hao J, Sawyer DB, Hatzopoulos AK, Hong CC (2011) Recent progress on chemical biology of pluripotent stem cell self-renewal, reprogramming and cardiomyogenesis. Rec Pat Regen Med 1:263–274

12. Lin T, Ambasudhan R, Yuan X, Li W, Hilcove S, Abujarour R, Lin X, Hahm HS, Hao E, Hayek A, Ding S (2009) A chemical platform for improved induction of human iPSCs. Nat Methods 6:805–808

13. Li W, Zhou H, Abujarour R, Zhu S, Young Joo J, Lin T, Hao E, Schöler HR, Hayek A, Ding S (2009) Generation of human-induced pluripotent stem cells in the absence of exogenous Sox2. Stem Cells 27:2992–3000

14. Hacein-Bey-Abina S, Von Kalle C, Schmidt M, McCormack MP, Wulffraat N, Leboulch P, Lim A, Osborne CS, Pawliuk R, Morillon E, Sorensen R, Forster A, Fraser P, Cohen JI, de Saint Basile G, Alexander I, Wintergerst U, Frebourg T, Aurias A, Stoppa-Lyonnet D, Romana S, Radford-Weiss I, Gross F, Valensi F, Delabesse E, Macintyre E, Sigaux F, Soulier J, Leiva LE, Wissler M, Prinz C, Rabbitts TH, Le Deist F, Fischer A, Cavazzana-Calvo M (2003) LMO2-associated clonal T cell proliferation in two patients after gene therapy for SCID-X1. Science 302:415–419

15. Woltjen K, Hämäläinen R, Kibschull M, Mileikovsky M, Nagy A (2011) Transgene-free production of pluripotent stem cells using piggy-Bac transposons. Methods Mol Biol 767:87–103

16. Yusa K, Rad R, Takeda J, Bradley A (2009) Generation of transgene-free induced pluripotent mouse stem cells by the piggyBac transposon. Nat Methods 6:363–369

17. Zhou H, Wu S, Joo JY, Zhu S, Han DW, Lin T, Trauger S, Bien G, Yao S, Zhu Y, Siuzdak G, Schöler HR, Duan L, Ding S (2009) Generation of induced pluripotent stem cells using recombinant proteins. Cell Stem Cell 4: 381–384

18. Kim D, Kim CH, Moon JI, Chung YG, Chang MY, Han BS, Ko S, Yang E, Cha KY, Lanza R, Kim KS (2009) Generation of human induced pluripotent stem cells by direct delivery of reprogramming proteins. Cell Stem Cell 4: 472–476

19. Zhang H, Ma Y, Gu J, Liao B, Li J, Wong J, Jin Y (2012) Reprogramming of somatic cells via TAT-mediated protein transduction of recombinant factors. Biomaterials 33: 5047–5055

20. Warren L, Manos PD, Ahfeldt T, Loh YH, Li H, Lau F, Ebina W, Mandal PK, Smith ZD, Meissner A, Daley GQ, Brack AS, Collins JJ, Cowan C, Schlaeger TM, Rossi DJ (2010) Highly efficient reprogramming to pluripotency and directed differentiation of human cells with synthetic modified mRNA. Cell Stem Cell 7:618–630

21. Tavernier G, Wolfrum K, Demeester J, De Smedt SC, Adjaye J, Rejman J (2012) Activation of pluripotency-associated genes in mouse embryonic fibroblasts by non-viral transfection with in vitro-derived mRNAs encoding Oct4, Sox2, Klf4 and cMyc. Biomaterials 33:412–417

22. Symons JA, Alcamí A, Smith GL (1995) Vaccinia virus encodes a soluble type I interferon receptor of novel structure and broad species specificity. Cell 81:551–560

23. Drews K, Tavernier G, Demeester J, Lehrach H, De Smedt SC, Rejman J, Adjaye J (2012) The cytotoxic and immunogenic hurdles associated with non-viral mRNA-mediated reprogramming of human fibroblasts. Biomaterials 33:4059–4068

24. Miyoshi N, Ishii H, Nagano H, Haraguchi N, Dewi DL, Kano Y, Nishikawa S, Tanemura M, Mimori K, Tanaka F, Saito T, Nishimura J, Takemasa I, Mizushima T, Ikeda M, Yamamoto H, Sekimoto M, Doki Y, Mori M (2011) Reprogramming of mouse and human cells to pluripotency using mature microRNAs. Cell Stem Cell 8:633–638

25. Lin SL, Chang DC, Chang-Lin S, Lin CH, Wu DT, Chen DT, Ying SY (2008) Mir-302 reprograms human skin cancer cells into a pluripotent ES-cell-like state. RNA 14:2115–2124

26. Lin SL, Chang DC, Lin CH, Ying SY, Leu D, Wu DT (2011) Regulation of somatic cell reprogramming through inducible mir-302 expression. Nucleic Acids Res 39:1054–1065

27. Lin SL (2011) Concise review: deciphering the mechanism behind induced pluripotent stem cell generation. Stem Cells 29: 1645–1649

Chapter 7

Transformation of *Bacillus subtilis*

Xiao-Zhou Zhang, Chun You, and Yi-Heng Percival Zhang

Abstract

Bacillus subtilis has tremendous applications in both academic research and industrial production. However, molecular cloning and transformation of *B. subtilis* are not as easy as those of *Escherichia coli*. Here we developed a simple protocol based on super-competent cells prepared from the recombinant *B. subtilis* strain SCK6 and multimeric plasmids generated by prolonged overlap extension-PCR. Super-competent *B. subtilis* SCK6 cells were prepared by overexpression of the competence master regulator ComK that was induced by adding xylose. This new protocol is simple (e.g., restriction enzyme, phosphatase, and ligase free), fast, and highly efficient (i.e., ~10^7 or ~10^4 transformants per μg of multimeric plasmid or ligated plasmid DNA, respectively). Shuttle vectors for *E. coli–B. subtilis* are not required.

Key words *Bacillus subtilis*, ComK, High-efficiency transformation, Molecular cloning, Multimeric plasmid, Super-competence

---

1 Introduction

Bacillus subtilis is the best known and most extensively studied model Gram-positive bacterium. Its derivative strains have tremendous applications in both academic research and industrial production. However, *B. subtilis* transformation is not as easy as that of *Escherichia coli*. The preparation of *B. subtilis* competent cells is usually a two-step procedure by using two types of minimal media [1]. For obtaining relatively high transformation efficiencies (e.g., ~10^3–10^4 transformants per μg of DNA), the cell growth must be monitored very carefully [1, 2]. The reported highest transformation efficiencies of *B. subtilis* are ~1–3×10^6 transformants per μg of DNA, by using either the two-step procedure with multimeric plasmids [3] or a high-osmolarity electroporation method [4]. However, both methods are labor intensive, tedious, and difficult to be operated by beginners.

To decrease the labor and achieve high *B. subtilis* transformation efficiency, we prepared super-competent *B. subtilis* cells that can be efficiently transformed by foreign DNA and multimeric plasmids [5].

Lianhong Sun and Wenying Shou (eds.), *Engineering and Analyzing Multicellular Systems: Methods and Protocols*,
Methods in Molecular Biology, vol. 1151, DOI 10.1007/978-1-4939-0554-6_7, © Springer Science+Business Media New York 2014

We modified *B. subtilis* 1A751 [6] by overexpressing its competence master regulator ComK [7], where the *comK* gene is controlled under the xylose-inducible promoter P_{xylA} [8]. The super-competent cells were prepared by adding xylose into the cells growing in the exponential phase for 2 h for inducing super-competence. Because *B. subtilis* prefers to be transformed by multimeric plasmids rather than monomeric ones [9], we generated multimeric plasmids by using prolonged overlap extension-PCR (POE-PCR) [10]. In addition, this protocol can be used to construct plasmids not only independent of restriction enzyme sites but also without the use of restriction enzymes and ligase. The utilization of the super-competent cells and multimeric plasmids for cloning greatly simplifies transformation and cloning in *B. subtilis* without relying on *E. coli*–*B. subtilis* shuttle vectors.

2   Materials

2.1 Biological and Chemical Materials

1. *Bacillus subtilis* strain SCK6 [5]: The genotype of SCK6 is *his, nprR2, nprE18, ΔaprA3, ΔeglS102, ΔbglT, bglSRV, lacA::PxylA-comK, erm*, which can be obtained from *Bacillus* Genetic Stock Center (http://www.bgsc.org) with an accession code of 1A976 (*see* **Note 1**).

2. Luria Broth (LB) liquid medium and 1.5 % (w/v) LB agar plates.

3. 20 % Xylose stock solution. Store at 4 °C.

4. 0.3 mg/mL erythromycin solution: Weigh 3 mg erythromycin and dissolve in 10 mL with absolute ethanol. Store at –20 °C.

5. 50 % glycerol: Take 50 mL 100 % glycerol; make up to 100 mL with deionized water. Store at room temperature.

6. DNA templates for plasmid backbone and inserted gene of interest.

7. New England Biolabs high-fidelity Phusion DNA polymerase (*see* **Note 2**).

8. Agarose.

9. Appropriate PCR primers (two pairs).

10. DNA purification kit for PCR product.

2.2 Equipment

1. Eppendorf thermocyler (model: Mastercycler ep gradient).

2. Agarose gel-running system.

3. Rotary shaking incubator.

3   Methods

3.1   Molecular Cloning by Generating Multimeric Plasmids

1. Two pairs of primers (IF, IR; and VF, VR) are needed to amplify the DNA fragments of insert gene of interest and vector backbone, respectively (Fig. 1). VF, the forward primer for vector linearization with a length of 50 bp, contains the last 25 bp of 3′ terminal of insert sequence and the first 25 bp of 5′ terminal of vector sequence. IF, the forward primer for amplifying insert with a length of 50 bp, contains the last 25 bp of 3′ terminal of vector sequence and first 25 bp of 5′ terminal of insert sequence. IR and VR are the reverse complementary sequences of VF and IF, respectively. Thus, IF and IR were used to amplify insert fragment; VF and VR were used to amplify vector backbone (*see* **Note 3**).

2. To prepare multimeric plasmids, the insert fragment was amplified with a pair of primers of IF and IR by using the NEB Phusion DNA polymerase. Similarly, the vector backbone was amplified with a pair of primers of VF and VR by using Phusion DNA polymerase (Fig. 2). The PCR reaction system contains dNTP, 0.2 mM for each; primers, 0.02 µM; template, 0.05 ng/µL; and Phusion DNA polymerase, 0.04 U/µL. The PCR program is 98 °C denaturation, 30 s; 30 cycles of 98 °C denaturation, 10 s; 60 °C annealing, 10 s; and extension at 72 °C at 3 kb/min for the targeted fragment. The PCR products were cleaned with a PCR DNA clean kit.

   The multimerization process (Fig. 2) was conducted based on the previous two PCR products without primers by using the Phusion DNA polymerase. The PCR conditions were 98 °C denaturation, 30 s and 30 cycles of 98 °C denaturation,

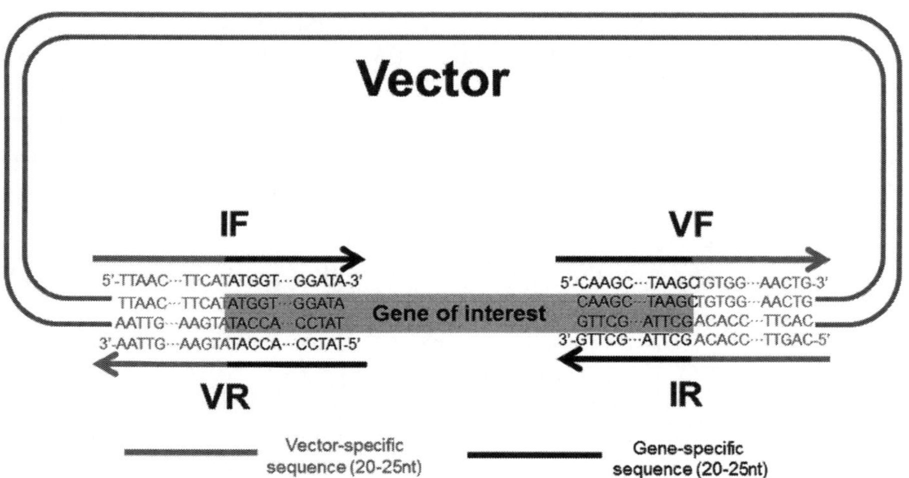

Fig. 1 Primer design for generating multimeric DNA

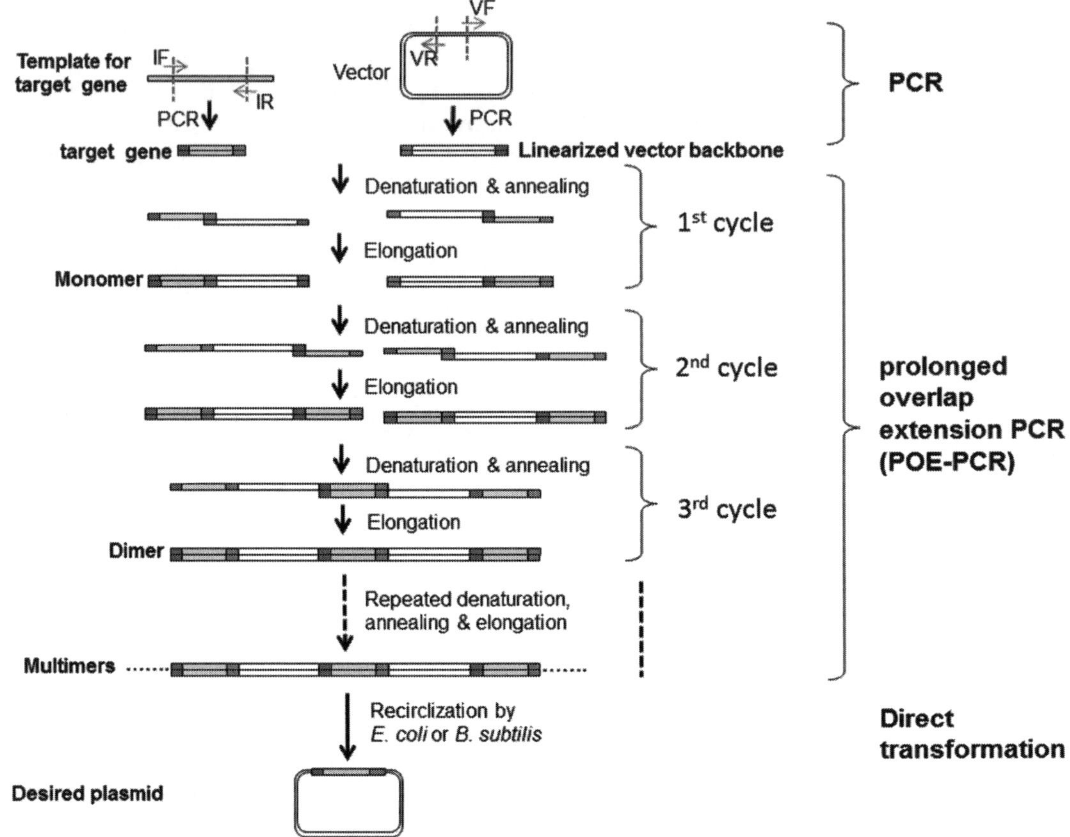

Fig. 2 The scheme of molecular cloning based on DNA multimers [10]. First, two 3′ and 5′ overlapped insertion and vector fragments are generated by regular PCR. Second, DNA multimers are formed by modified overlap extension. Third, *E. coli* or *B. subtilis* strains can cleave DNA multimers to a circular plasmid, a desired chimeric plasmid

10 s; 60 °C annealing, 10 s; and extension at 72 °C for 2 kb/ min for the length of the resulting plasmid. This PCR reaction system contained dNTP, 0.2 mM for each; insert fragment, 2 ng/μL; vector backbone, equimolar with insert fragment; and Phusion polymerase, 0.04 U/μL. Compared to regular PCR, 1.5–2-fold longer extension time was used, the so-called POE-PCR.

3. The POE-PCR product—multimeric plasmids—can be examined by 0.8 % agarose gel (Fig. 3). Approximately 10–100 ng/ μL multimeric plasmids were usually obtained. The multimeric plasmids with branched structure were high-molecular-weight products so that they cannot migrate into the gel (Fig. 3, lane 3). To check whether the insert was incorporated into the plasmid backbone in POE-PCR, the PCR product can be digested by restriction enzymes, as shown in Fig. 3, lane 4 (*see* **Note 4**).

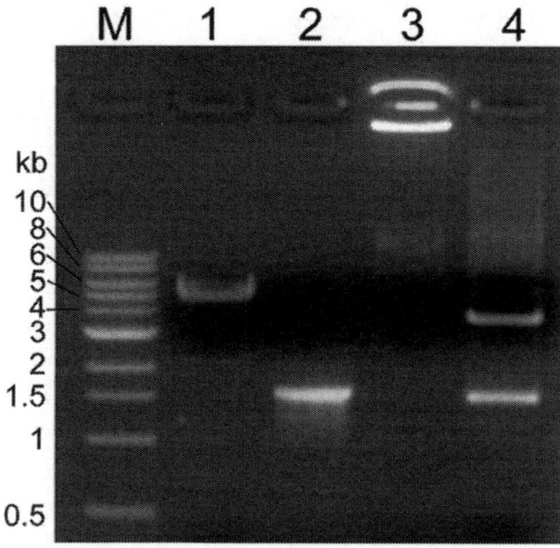

Fig. 3 Generation of recombinant multimeric plasmids by overlap extension PCR. *Lanes*: *M*, DNA markers; *1*, PCR-linearized vector backbone; *2*, PCR product of the insert; *3*, multimeric plasmids after overlap extension PCR; *4*, multimeric plasmids digested with restriction enzymes for verification purpose

3.2 Preparation and Transformation of B. subtilis Super-Competence Cells

1. To revive the strain from −80 °C stock, a loop of *B. subtilis* strain SCK6 from −80 °C stock is obtained and subsequently streaked on the LB plate supplemented with 0.3 μg/mL erythromycin. Incubate the plate at 37 °C overnight.

2. To prepare the seed culture, a single colony is inoculated from the plate into 50 mL LB medium containing 0.3 μg/mL erythromycin in a 250-mL flask. Incubate cell culture at 37 °C in a rotary shaking incubator for 8–12 h.

3. To induce super-competence, the absorbency of the seed culture at 600 nm is measured and the culture is diluted to $A_{600} = \sim1.0$ by adding freshly prepared ~37 °C LB medium containing 0.3 μg erythromycin/mL. Add D-xylose into the diluted culture at the final concentration of 1 % (w/v). Incubate the diluted cell culture in rotary shaking incubator for another 2 h, and then the cells are ready for direct transformation (*see* **Note 5**).

4. Mix 1–2 μL of the POE-PCR product with 100 μL of competent cells in a cell tube. Incubate cells in a rotary shaking incubator at 37 °C at 200 rpm for 1.5 h (*see* **Note 6**) to complete transformation.

5. Make serial dilutions of transformed competent cells, and spread these dilutions on LB plate with the appropriate antibiotic. Incubate the plates at 37 °C for 8–12 h (*see* **Note 7**) to select transformants.

6. To verify selected transformants, they are inoculated from the plate into LB medium with appropriate antibiotic and incubated

at 37 °C for 8–10 h. Centrifuge the cell culture to collect the cells and extract recombinant plasmids for sequencing or restriction enzyme digestion (*see* **Note 8**).

7. Gently mix the competent cells with certain amount of 50 % glycerol to make the final concentration of glycerol to be 15 %, and then stock the competent cells at −80 °C for future transformation (*see* **Note 9**).

4   Notes

1. *B. subtilis* 168 and any of its derivative with an intact *lacA* site can be converted to a recombinant strain with a super-competence feature by the double-crossover integration of linearized vector pAX01-comK (BGSC accession code: ECE222) into the chromosome for xylose-induced expression of ComK. Also, multimeric plasmids generated by POE-PCR can be applied to most *B. subtilis* strains with higher transformation efficiency as compared to those of regular shuttle vector transformation. In addition, the induced super-competence by overexpressing the *comK* gene could be applied to other *Bacillus* species.

2. High-fidelity DNA polymerase is required here to avoid possible mutations during the PCR process. The high-fidelity DNA polymerase should also have good performance for amplifying long DNA fragments.

3. We usually use the overlap region of 25 bp; that is, the whole primer length is 50 bp. Based on our experience, the overlap region should be at least ~20 bp for good performance of the overlap extension process. The overlap region may be longer, but longer primer could result in high possibility to get mutations in the synthesized primers and the unit price for primer could be much higher when its length exceeds 60 bp.

4. It is normal to see some high-molecular-weight DNA at the position of more than 10 kb, which are some intermediates generated during the multimerization process. The POE-PCR product, a high-viscosity solution, can be used to transform competent cells directly. Do not clean the PCR product with a DNA cleaning kit. The PCR product of multimeric plasmids can be stored at 4 °C for at least a week. Do not freeze the PCR product since it may cause DNA precipitation.

5. To avoid the drastic change in cell culture conditions, the freshly added LB medium should be pre-warmed to 37 °C before diluting the seed culture.

6. Do not add too much PCR product in competent cells for transformation, since the PCR solution decreases the

transformation efficiency [3]. Usually, we can get ~10^7 transformants per μg of the POE-PCR product (i.e., multimeric plasmids).

7. Making serial dilutions before spreading on the plate is necessary, especially for beginners, because transformation efficiencies may vary greatly.

8. To get the good-quality plasmids, do not extract it from overnight cell culture. Plasmids could be severely contaminated by genomic DNA if the cell culture was too old.

9. The competent cells stored at −80 °C can be used for at least 1 year. If transformation efficiency was decreased dramatically, new competent cells should be prepared from the original stock strain.

Acknowledgement

This work was supported by the DOE BioEnergy Science Center and the College of Agriculture and Life Sciences Biodesign and Bioprocessing Research Center at Virginia Tech to Y.P.Z. X.Z.Z. appreciates the support from NSF and DOE SBIR grants.

References

1. Cutting SM, Vander Horn PB (eds) (1990) Molecular biological methods for *Bacillus*. Wiley, Chichester

2. Ehrlich SD (1978) DNA cloning in *Bacillus subtilis*. Proc Natl Acad Sci USA 75(3):1433–1436

3. Shafikhani S, Siegel RA, Ferrari E, Schellenberger V (1997) Generation of large libraries of random mutants in *Bacillus subtilis* by PCR-based plasmid multimerization. Biotechniques 23(2):304–310

4. Xue G-P, Johnson JS, Dalrymple BP (1999) High osmolarity improves the electro-transformation efficiency of the gram-positive bacteria *Bacillus subtilis* and *Bacillus licheniformis*. J Microbiol Methods 34(3):183–191

5. Zhang XZ, Zhang YHP (2011) Simple, fast and high-efficiency transformation system for directed evolution of cellulase in *Bacillus subtilis*. Microb Biotechnol 4(1):98–105

6. Wolf M, Geczi A, Simon O, Borriss R (1995) Genes encoding xylan and beta-glucan hydrolysing enzymes in *Bacillus subtilis*: characterization, mapping and construction of strains deficient in lichenase, cellulase and xylanase. Microbiology 141(Pt 2):281–290

7. Susanna KA, Fusetti F, Thunnissen A-MWH, Hamoen LW, Kuipers OP (2006) Functional analysis of the competence transcription factor ComK of *Bacillus subtilis* by characterization of truncation variants. Microbiology 152(2): 473–483. doi:10.1099/mic.0.28357-0

8. Hartl B, Wehrl W, Wiegert T, Homuth G, Schumann W (2001) Development of a new integration site within the *Bacillus subtilis* chromosome and construction of compatible expression cassettes. J Bacteriol 183(8):2696–2699. doi:10.1128/JB.183.8.2696-2699.2001

9. Canosi U, Morelli G, Trautner TA (1978) The relationship between molecular structure and transformation efficiency of some *S. aureus* plasmids isolated from *B. subtilis*. Mol Gen Genet 166(3):259–267

10. You C, Zhang XZ, Zhang YH (2012) Simple cloning via direct transformation of PCR product (DNA multimer) to *Escherichia coli* and *Bacillus subtilis*. Appl Environ Microbiol 78(5): 1593–1595. doi:10.1128/AEM.07105-11

Chapter 8

Culturing Anaerobes to Use as a Model System for Studying the Evolution of Syntrophic Mutualism

Sujung Lim, Sergey Stolyar, and Kristina Hillesland

Abstract

Our current understanding of the evolution of mutualisms is limited partly because there have been relatively few model systems for studying it in real time. A model mutualistic interaction between the bacterium *D. vulgaris* and the archaeaon *M. maripaludis* was developed to allow for rigorous tests of general hypotheses about the evolution and ecology of mutualisms. This model system also allows us to develop an evolutionary genetics perspective on an interaction that plays a key ecological role in many oxygen-free microbial communities. Here, we describe the techniques used to make anoxic media for propagating these species alone or in conditions that require their cooperation.

Key words Syntrophy, Mutualism, Sulfate-reducing bacteria, Methanogen, Anaerobic techniques, Widdel flask

1 Introduction

Mutually beneficial interactions between species have long been of interest because of their critical roles in the functioning of ecosystems (e.g., symbiotic nitrogen fixation [1], coral-algae symbioses [2], and gut microbial communities [3]) and because of the evolutionary and physiological puzzles associated with cooperation between species [4–6]. However, until recently, few experimental microbial models of mutualism have been developed [7–10]. To understand how cooperation between species affects their evolution, stability, and population dynamics, we developed a mutualism that is genetically tractable, that can be readily propagated and controlled in the laboratory, and where both species can be studied together or in isolation [11, 12]. To maximize the impact of experiments testing general hypotheses about mutualisms and how they evolve, we use one that is similar to a naturally occurring and widespread interaction that plays a key ecological role in many oxygen-free environments. The model system, its ecological relevance, and the methods used to culture it are described below.

Lianhong Sun and Wenying Shou (eds.), *Engineering and Analyzing Multicellular Systems: Methods and Protocols*,
Methods in Molecular Biology, vol. 1151, DOI 10.1007/978-1-4939-0554-6_8, © Springer Science+Business Media New York 2014

1.1 An Anaerobic Microbial Mutualism

Cooperative interactions between microbial species play a crucial role in carbon decomposition in many oxygen-free environments. In the absence of oxygen, decaying polymeric substrates such as polysaccharides, proteins, and lipids are broken down into progressively smaller organic molecules, hydrogen, and carbon dioxide by a community of microorganisms that feed off of one another's by-products [13–15]. Microbial species are able to gain energy from the fermentation of these organic molecules and by coupling this breakdown of organic molecules with the reduction of sulfate, nitrate, and other molecules that can serve as electron acceptors. However, in environments such as lake sediments or the rumen of cows, these electron acceptors are often unavailable and fermentation alone cannot always provide enough energy for growth. In such conditions, the final steps of carbon decomposition are fueled by cooperation between fermenters and species that consume hydrogen and other products of these reactions (i.e., formate, acetate). This cooperation is necessary because the first step in the fermentation of carbon sources such as lactate or propionate is thermodynamically unfavorable in standard conditions. However, when the products of such reactions are maintained at a very low concentration, they can generate more energy, enabling the fermenters to grow and continue breaking down carbon while at the same time providing the hydrogen consumers with food [13–15]. Such cooperative interactions between microbial species are called syntrophy.

Syntrophic interactions between hydrogenotrophic, methanogenic archaea and fermentors are responsible for the reduction of oxidized carbon to methane in rice paddies, rumen, lake sediments, and anaerobic digesters used to process human waste [13–15]. These interactions were first discovered in 1967 by Bryant et al. [16] when they isolated a bacterium from sewage sludge that could ferment ethanol, producing methane as a by-product. They found that the culture actually had two species. Since then, the biochemistry, physiology, and ecology of syntrophs have been studied extensively in order to understand how anaerobic communities function and how microorganisms can survive using very small changes in free energy [15]. Other syntrophy models with *Desulfovibrio vulgaris* were developed to study its physiology, although some inferences about its evolutionary history were made [17–19]. No one has systematically addressed how species evolve in response to selective pressure imposed by syntrophic cooperation.

To develop a deeper understanding of the origins and evolution of syntrophies—and mutually beneficial interactions in general, we (in coordination with Dave Stahl) [11, 12] constructed a syntrophy from two genetically tractable species that had no recent history of adaptation to syntrophy and no prior history of interaction with each other (Fig. 1). The availability of genomic

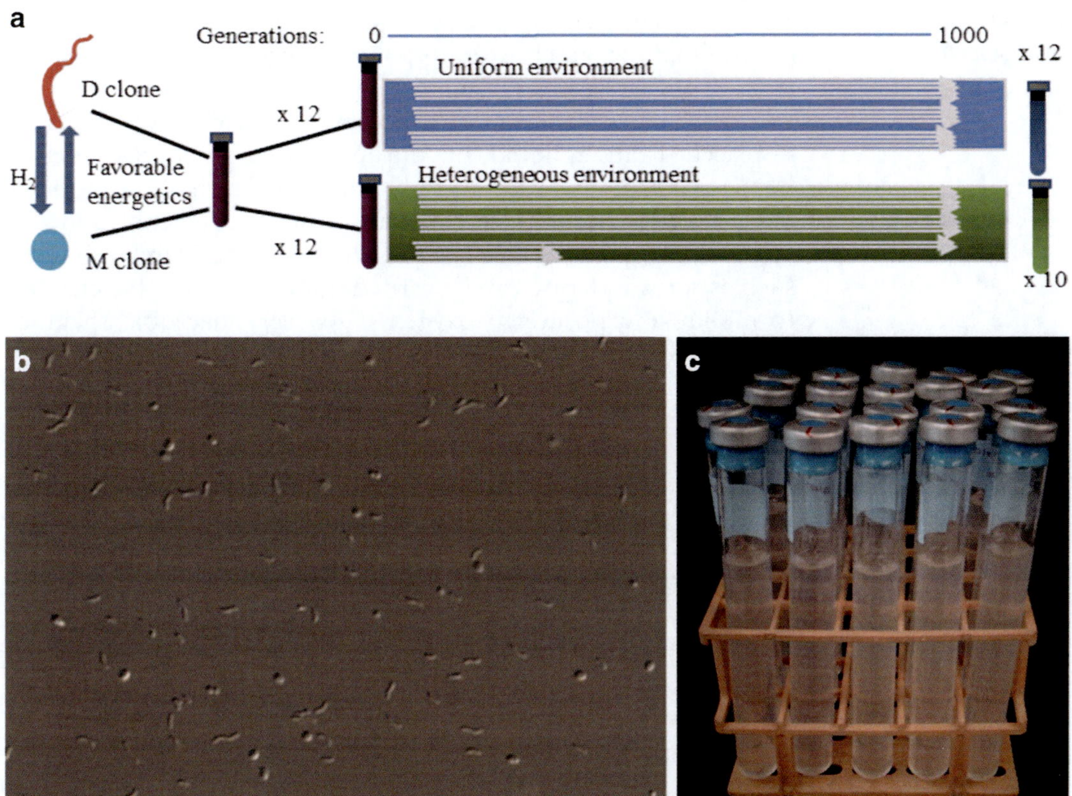

Fig. 1 (**a**) Diagram of the syntrophic mutualism and the evolution experiment. In the uniform environment (*light blue*), tubes are constantly shaking at 300 rpm in a horizontal position. Tubes were incubated upright and without shaking in the heterogeneous environment (*green*). (**b**) Coculture of *D. vulgaris* (*vibroid-shaped cells*) and *M. maripaludis* (*spherical cells*). (**c**) Balch tubes containing cocultures

sequences for both species has enabled studies of gene expression and metabolic modeling of the syntrophy [12, 20, 21]. In this interaction, the sulfate-reducing bacteria *D. vulgaris* [22] ferments lactate (the only carbon and energy source in the experimental conditions), producing acetate, hydrogen, and carbon dioxide as by-products. *Methanococcus maripaludis* gains energy by using this hydrogen to reduce carbon dioxide to methane, thereby keeping the concentration of hydrogen at a low level and enabling *D. vulgaris* to continue gaining energy from the fermentation of lactate [12]. After 300 generations of adaptation to syntrophy, all 22 independently evolved cocultures were significantly more stable than the ancestral cocultures and produced more cellular material per mole lactate at a higher rate than the ancestors [11]. Here, we describe the techniques used to culture these species together and separately.

1.2 Overview of Techniques for Studying Anaerobes

Successful cultivation of anaerobic syntrophies requires carefully controlling the gaseous atmosphere to exclude oxygen at each step. There are several different techniques for making oxygen-free media for the cultivation of anaerobes [23–25]. Oxygen can be displaced from a liquid by introducing another gas, such as nitrogen, through a cannula bubbling up from the bottom of the flask. Boiling the liquid media and allowing it to cool under an alternative gas can also remove dissolved oxygen. Many techniques such as spread-plating can be carried out in an anaerobic chamber to minimize exposure to oxygen. However, anaerobic chambers can be costly, consume valuable laboratory space, and accumulate higher concentrations of potential contaminating microorganisms. Thus, scientists have developed several techniques for propagating anaerobes without using anaerobic chambers (e.g., refs. 23, 24).

We use the procedure of Widdel et al. [25] to make anaerobic media for propagating *D. vulgaris* and *M. maripaludis* either in monocultures or together in syntrophic cocultures without using an anaerobic chamber. In this technique, the media is autoclaved and dispensed with a specially made inverted flask. The flask is sealed with a rubber stopper with ports for gas exchange and for dispensing media. Anoxic gas flows through the headspace of the flask as the media cools and is dispensed. Screw caps on the top of the flask can be loosened to let gas flow out of the vessel or tightened to pressurize the vessel, pushing the liquid media out through the dispensing tube into sterile Balch tubes. A clamp is used to start and stop the flow of media. This procedure allows the researcher to make large batches of media that require several post-autoclave amendments without using an anaerobic chamber.

The formulation of the media described below, named CCM, was developed by modifying media for growth of *Desulfovibrio* [25] and for growth of marine halotolerant *Methanococcus* [26]. Because *D. vulgaris* is inhibited by the high salt concentration that was present in the *M. maripaludis* medium, and *M. maripaludis* requires magnesium ions for methanogenesis, we decreased the amount of sodium chloride to 2.28 g/L and kept $MgCl_2$ at a relatively high level [12].

While *D. vulgaris* is able to survive in the presence of some oxygen [27], many methanogens, including *M. maripaludis*, are very sensitive to even short periods of exposure to oxygen [28]. However, simply removing all oxygen from the environment is not enough to achieve optimum culture conditions for these species. The environment must also be kept at a very low redox potential to maximize the possible energy gain for both species [25, 28, 29]. To achieve this, sodium sulfide and cysteine are added to the culture media [23–25, 28].

Below, we describe our procedure for making media with this technique, how to make and transfer anaerobic liquids and cultures, and how the media recipe can be varied to grow *D. vulgaris* and *M. maripaludis* in coculture or monoculture.

2   Materials

All solutions must be prepared with ultrapure type I water. Stock solutions must be made sterile and anaerobic. Anaerobic stock solutions are made by boiling water with N_2 flowing across the top of the surface and then adding solutes after the water cools. Dispense solutions into anaerobic serum bottles (e.g., Wheaton, part no. W012465), cap with a rubber stopper and aluminum seal, crimp, and then autoclave. After autoclaving, flush headspace with sterile N_2 (*see* **Note 1**), except when noted otherwise.

2.1   For Transferring Liquids and Cultures Anaerobically

1. Two prepared anaerobic solutions, one that you are transferring from and another that you are transferring to. If you are transferring cultures, you need one dense culture and a tube of fresh media.

2. Gassing setup: Connect a N_2 gassing line to a sterilized glass syringe filled with glass wool, and attach a bent cannula to the end. A sterile syringe filter in place of the glass syringe would also work; the objective is to ensure that the gas exiting the cannula is sterile.

3. Sterile syringes and 23 G 1″ needles (BD, part no. 305193).

2.2   For Dispensing Media into Tubes

1. Anaerobic stocks of 1 M K_2HPO_4, 1 M $NaHCO_3$ (*see* **Note 2**), 1 M L-cysteine hydrochloride, 7.8 % w/v $Na_2S \cdot 9H_2O$ (*see* **Note 3**).

2. Anaerobic stock of 1,000× trace minerals: Trace minerals contain, per liter, 12.8 g nitrilotriacetic acid, pH 6.5; 1.0 g $FeCl_2 \cdot 4H_2O$; 0.5 g $MnCl_2 \cdot 4H_2O$; 0.3 g $CoCl_2 \cdot 6H_2O$; 0.2 g $ZnCl_2$; 0.05 g $Na_2MoO_4 \cdot 2H_2O$; 0.02 g H_3BO_3; 0.09 g $NiSO_4 \cdot 6H_2O$; 0.002 g $CuCl_2 \cdot 2H_2O$; 0.006 g $Na_2SeO_3 \cdot 5H_2O$; and 0.008 g $Na_2WO_4 \cdot 2H_2O$. The solution is autoclavable.

3. Anaerobic stock of 1,000× Thauer's vitamins (*see* **Note 4**): Thauer's vitamins contain, per liter, 0.02 g biotin; 0.02 g folic acid; 0.1 g pyridoxine HCl; 0.05 g thiamine HCl; 0.05 g riboflavin; 0.05 g nicotinic acid; 0.05 g DL-pantothenic acid; 0.05 g *p*-aminobenzoic acid; and 0.01 g cyanocobalamin (vitamin B_{12}).

4. CCMA (coculture medium A as per ref. 12) base solution: The base solution contains, per liter, 2.28 g NaCl; 5.5 g $MgCl_2 \cdot 6H_2O$; 0.14 g $CaCl_2 \cdot 2H_2O$; 0.5 g NH_4Cl; 0.1 g KCl; 5.6 g Na DL-lactate (60 %, syrup, Sigma); and 1 mL 0.1 % w/v resazurin (*see* **Note 5**), prepared in a flask. The pH should be adjusted to 7.2 with 5 M NaOH or HCl, as appropriate.

5. Gassing setup from Subheading 2.1.

6. Widdel flask: *See* Fig. 2 for a description of the Widdel flask. It can be specially made from a standard Pyrex flask at a glass shop. The stopper is made of butyl rubber. The ports in the stopper

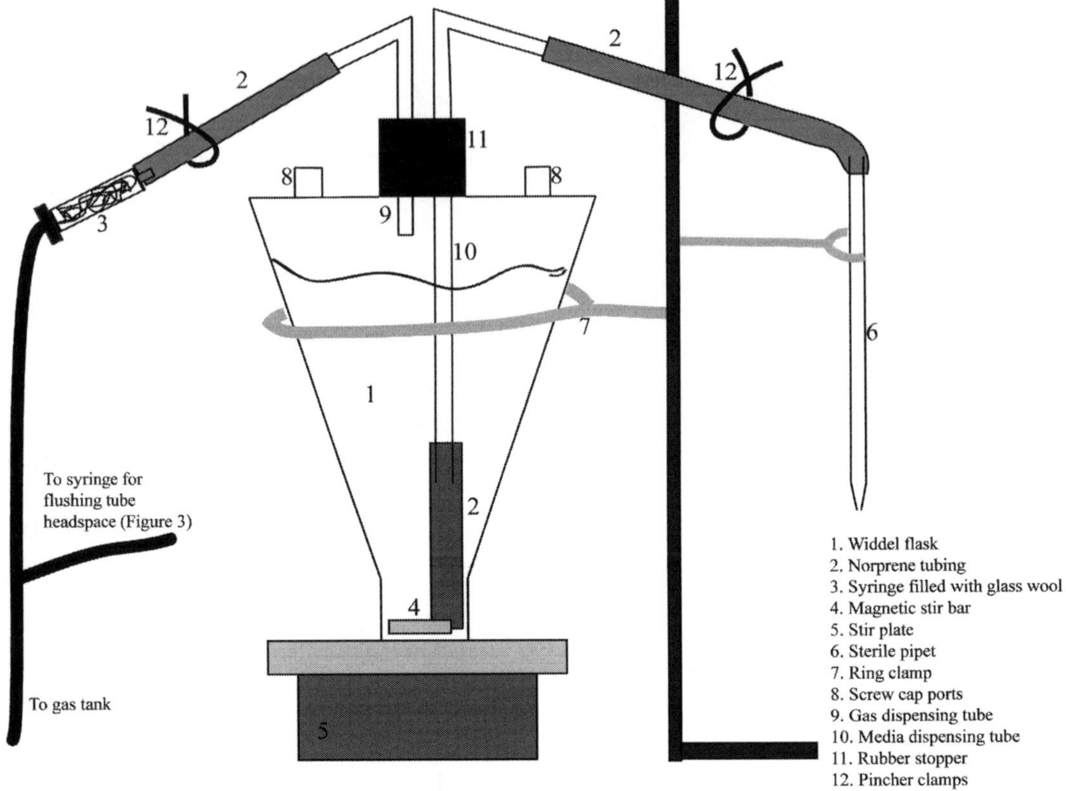

Fig. 2 Widdel flask and accessories for making anaerobic media

consist of glass rods connected to flexible Norprene tubing (Norprene formulation A-60-G, 1/4″ ID × 3/8″ OD).

7. Aluminum seals (Wheaton, part no. 224178) and a crimper for installing the aluminum seals.

8. Tweezers.

9. Sterile Balch tubes (Fig. 1; Bellco, part no. 2048-00150) and a beaker filled with sterile Balch tube stoppers (Bellco, part no. 2048-1180).

2.3 For Anaerobic Growth of D. vulgaris, M. maripaludis, and the Coculture

1. Balch tubes containing appropriate media (Table 1).

2. 37 °C incubator.

3. Compressed gas cylinders of 80 % N_2:20 % CO_2 or 80 % H_2:20 % CO_2 depending on which culture is to be propagated.

Table 1
Growth requirements for cultivating both species as a monoculture or a coculture

	D. vulgaris	*M. maripaludis*	Syntrophic coculture
Electron donor	60 % Na DL-lactate, 5.6 g/L	Hydrogen	60 % Na DL-lactate, 5.6 g/L
Electron acceptor	Na_2SO_4, 4.3 g/L or $Na_2S_2O_3 \cdot 5H_2O$, 1.24 g/L	Carbon dioxide	No added chemical electron acceptor
Volume per tube	10 mL	5 mL	20 mL
Headspace pressure and chemical composition	80 % N_2:20 % CO_2, atmospheric pressure	80 % H_2:20 % CO_2, 275.8 kPa (gauge)	80 % N_2:20 % CO_2, atmospheric pressure
Miscellaneous notes	Na_2SO_4 is added pre-autoclave; $Na_2S_2O_3$ is added post-autoclave from a sterile anaerobic 1 M stock solution, after Na_2S has been added and the medium has fully reduced	*Carbon source:* Sodium acetate (anhydrous), 0.82 g/L *Amino acids:* Casamino acids, 1 g/L *Reductant:* Add 3 mL/ L 1 M cysteine solution instead of 1 mL/L, Omit Na_2S until prior to inoculation—add 0.1 mL 2.5 % w/v solution *Omit* Na DL-lactate	

3 Methods

3.1 Transferring Solutions and Cultures Anaerobically

1. Sterilize the metal gassing cannula by flaming it and the tops of the anaerobic bottles by flooding the tops with ethanol and burning it off with a flame.

2. Turn on the N_2 cylinder, and hold a wet hand up to the gassing cannula to make sure that gas is running through it.

3. Remove residual oxygen from the needle and syringe (to be used for transferring liquids later) by inserting the needle into the gassing cannula. Alternately depress and then draw out the plunger for 5–10 s to turn the barrel of the syringe anaerobic.

4. Draw out an approximately equal volume of gas as the volume of liquid being removed to maintain nearly neutral pressure in the sealed bottles or tubes.

5. Immediately insert the needle and syringe into the bottle from which liquid will be removed. Push the gas into the bottle.

6. Turn the bottle or tube upside down with the needle and syringe still pushed through the stopper. Pull on the syringe to remove the desired volume of liquid. Then carefully pull out the needle and syringe. Make sure that there are not any bubbles in the syringe (*see* **Note 6**).

7. Insert the needle and syringe into the anaerobic culture or solution that you are transferring to. Depress the plunger, and remove and discard the needle and syringe in an appropriate sharps container. If you are transferring more than 1 mL, be sure to remove an equal volume of gas from the container to keep the pressure constant (*see* **Note 7**).

3.2 Anaerobic Preparation of Tubes for Culturing D. vulgaris and M. maripaludis (See Note 8)

1. Pour the CCMA base solution into the Widdel flask along with an appropriately sized stir bar. Seal the Widdel flask with the stopper and septated screw caps (*see* Fig. 2). All tubing leading out of the stopper must be clamped shut. Wrap the exposed ends of the tubing in aluminum foil to maintain sterility after the flask is autoclaved. Make sure that the screw caps are attached but loose to allow steam to vent during autoclaving.

2. Autoclave the CCMA base solution in the Widdel flask (Fig. 2) at 121 °C for 30 min. Immediately remove the flask from the autoclave when sterilization is completed. Stabilize the flask with a ring stand. Use a needle to connect the 80:20 N_2:CO_2 line to the syringe with glass wool, as depicted in Fig. 2. Open the clamp on the tubing between the flask and the gas line to allow gas to flow into the Widdel flask. Close the screw caps, but leave one of them slightly loose so that gas can flow out of the Widdel flask. Cool the media by putting a bucket of ice water under the flask. Use the stir plate to gently stir the solution (*see* **Note 9**).

3. When the medium is at room temperature, add 1.1 mL/L 1 M K_2HPO_4, 25 mL/L 1 M $NaHCO_3$, 1 mL/L trace minerals, 1 mL/L Thauer's vitamins, 1 mL/L 1 M L-cysteine monohydrochloride, and 1 mL/L 7.8 % w/v sodium sulfide, as described in Subheading 2.1. Add them in the order specified through the ethanol-sterilized septum of the screw caps. As you add the solutions, keep them anaerobic. Turn off stirring while you add these stocks to minimize penetration of oxygen into the media. The media will initially be blue, gradually turn to pink, and then become colorless as it reduces. Once you have stopped adding solutions and closed the caps, you can start stirring the solution again.

4. After the medium has finished reducing and is no longer pink, aseptically remove the cotton plug from the end of a sterile 5 mL pipette and attach pipette to the media outlet tubing.

5. Slowly tighten septated caps on Widdel flask until sufficient positive pressure accumulates within the volume of the flask to

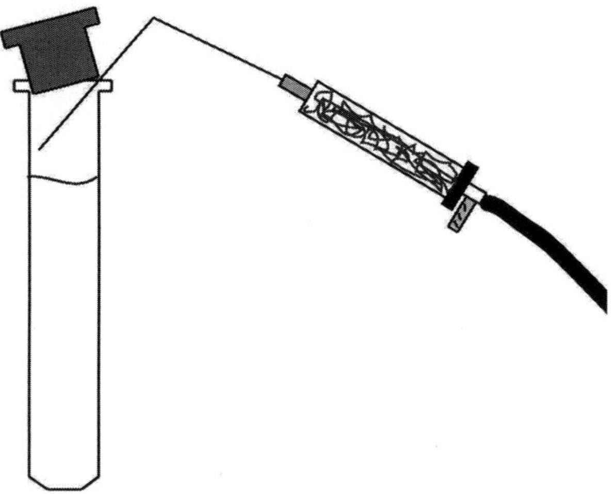

Fig. 3 Procedure for gassing headspace of a Balch tube. A syringe filled with glass wool is attached to the gas line to sterilize the gas before it enters the vessel. A toggle can be used to quickly turn the gas flow up or down. The cannula attached to the syringe rests on the side of the Balch tube and blows mixed gas across the surface of the liquid. The cannula is pulled out just as the stopper seals the tube

cause media to fill the outlet tube. Open pinch clamp to dispense media (*see* **Note 10**). Discard the first 20–30 mL of media. You are now ready to dispense media into tubes.

6. Hold a sterile 18 × 150 mm Balch tube (Bellco Glass) over the pipet attached to the Widdel flask so that the tip of the pipet touches the bottom of the tube. Open the clamp. As the media enters the tube, move the Balch tube down so that the tip remains just under the surface of the liquid (*see* **Note 11**). Close the clamp when the tube is about ¾ full. This is equivalent to 20 mL (*see* **Note 12**).

7. Fill the headspace of the tube with N_2:CO_2 by putting a gassing cannula in the tube. The gassing cannula should have been sterilized by passing through a flame, and the tip of the cannula should remain above the liquid (Fig. 3 and *see* **Note 13**). Gas the filled tube for about 20 s.

8. Seal tube with sterile 20 mm butyl rubber stoppers while maintaining gas flow (*see* **Note 14**). To avoid getting oxygen into the headspace, pull out the cannula just as the stopper seals the top of the tube. Tubes may turn pink during this process, but they will eventually become colorless again unless oxygen gets into the tube during the dispensing or sealing process. Make sure that the stoppers are completely inserted into the tubes. The stoppers may be twisted to facilitate pushing them all the way in.

9. Repeat **steps 6–8** until all of the media has been dispensed (*see* **Note 15**).

10. Use the capper to crimp an aluminum cap onto each tube (*see* **Notes 16, 17**). The aluminum seal will keep the stoppers from popping out of the tube.

3.3 Culturing D. vulgaris and M. maripaludis Independently and in Syntrophic Coculture

1. Follow the methods described in Subheadings 3.1 and 3.2 to make anaerobic tubes and inoculate them. Use Table 1 to determine how to modify the media or the gas phase for the species of interest.

2. Cultures should be incubated at 37 °C static and upright or shaking at 300 rpm in a horizontal position depending on the experimental objective.

3.3.1 Confirmation of Coculture Purity

Purity of cocultures and monocultures should be regularly monitored by microscopy. *M. maripaludis* is a pleomorphic coccus with dimensions averaging 1.2×1.6 μm [29] but can swell under conditions of stress up to 2.5 μm. *D. vulgaris* is a vibroid rod 0.4–0.8×1.5–4.0 μm long [30, 31]. Microscopic entities not fitting either of these characteristics should be regarded with caution as they may be undesirable, especially if the coculture has erratic growth or if there are dramatic changes in carrying capacity.

In addition, culture-based methods can be used to test for the presence of some contaminants. Cultures can be plated on heterotrophic media such as nutrient agar or tryptic soy agar and incubated aerobically and anaerobically. Neither *D. vulgaris* nor *M. maripaludis* should form dense colonies on these media, but common contaminants will. Some *Desulfovibrio* strains may form a thin film.

4 Notes

1. To gas the headspace of a sealed bottle, insert a needle (venting needle) into the bottle to let gas escape as new gas is added. Insert gas into the bottle by attaching a needle and sterile filter to the gas line. Insert the gassing needle into the bottle and flush. Take the gas line out before the venting needle to avoid pressure buildup in the bottle. Alternatively, if you want the bottle to be pressurized, remove the venting needle before pulling out the gassing needle.

2. Sodium bicarbonate must be stored with 80 %:20 % N_2:CO_2 in the headspace instead of nitrogen.

3. Sodium sulfide is toxic, and this solution should be made in a fume hood. It also quickly oxidizes in the presence of oxygen. Thus, it is especially important to make sodium sulfide stocks with anaerobic water.

4. Thauer's vitamin solution is not autoclavable; filter-sterilize it (0.2 µm filter size) into sterile bottles and cap with sterile stoppers. Gas headspace with sterile N_2, and store solution at 4 °C.

5. Resazurin is an indicator of both pH and redox potential. When the media is completely reduced and at a neutral pH, the solution will be colorless. If too much oxygen penetrates a tube, media containing resazurin will turn pink.

6. Bubbles may form in the syringe. To remove the bubbles, flush the syringe with liquid, or try tapping the syringe to dislodge the bubble and then push it back out into the bottle.

7. To achieve a more accurate volume measurement from the syringe, pull out more volume than you need, dispense the correct amount, and leave 0.1 mL or more in the syringe.

8. This entire procedure takes about 5 h, depending on how quickly the autoclave warms up and cools down. Once the media has come out of the autoclave, it cannot easily be saved and kept anaerobic in the Widdel flask for long periods. Thus, it is best to make sure that this procedure is not started unless there is enough time to complete the whole thing without stopping. The CCMA base solution can be made a day early and kept in a refrigerator overnight if needed.

9. To confirm that the gas is indeed running through the syringe and across the headspace of the Widdel flask, open one of the screw caps a little and listen for the sound of the liquid in the lid bubbling.

10. When sufficient pressure has accumulated in the Widdel flask, the liquid will begin to push out of the tube that reaches all the way to the bottom.

11. The tip must be held under the liquid to ensure that the new media pouring in does not contact the air and become oxygenated. The liquid will become oxygenated if the tip is kept at the bottom because the liquid will push up into the oxygen. If the tip is above the liquid surface, the media will flow through the air and become oxygenated.

12. You can also mark all of your tubes at a height that is equivalent to 20 mL to make it easier to ensure that you have the same volume in each tube.

13. If the gas flows too fast on the top of the liquid, it can cause the media in the tube to splatter. Adjust the airflow accordingly.

14. Keep the stoppers sterile by picking them up at the top only with tweezers that have been sterilized by flaming.

15. To save time, you can fill a second tube while the first tube is gassing.

16. Aluminum caps can be added the next day if necessary.

17. Different colors of aluminum caps can be used to differentiate between media types.

Acknowledgements

This material is based upon the work supported by the National Science Foundation under Grant No. DEB-1257525. Any opinions, findings, and conclusions or recommendations expressed in this material are those of the author(s) and do not necessarily reflect the views of the National Science Foundation. This work was also supported in part by ENIGMA—Ecosystems and Networks Integrated with Genes and Molecular Assemblies (http://enigma. lbl.gov), a Scientific Focus Area Program at Lawrence Berkeley National Laboratory, and the Office of Science, Office of Biological and Environmental Research, of the US Department of Energy under Contract No. DE-AC02-05CH11231.

References

1. Herridge DF, Peoples MB, Boddey RM (2008) Global inputs of biological nitrogen fixation in agricultural systems. Plant Soil 311:1–18

2. Baker AC (2003) Flexibility and specificity in coral-algal symbiosis: diversity, ecology, and biogeography of *Symbiodinium*. Annu Rev Ecol Evol Syst 34:661–689

3. Mackie RI (2002) Mutualistic fermentative digestion in the gastrointestinal tract: diversity and evolution. Integr Comp Biol 42:319–326

4. Bergstrom CT, Bronstein JL, Bshary R, Connor RC, Daly M, Frank SA et al (2003) Group report: interspecific mutualism – puzzles and predictions. In: Hammerstein P (ed) Genetic and cultural evolution of cooperation. MIT Press, Cambridge, pp 241–256

5. Sachs JL, Mueller UG, Wilcox TP, Bull JJ (2004) The evolution of cooperation. Q Rev Biol 79:135–160

6. Jones EI, Bronstein JL, Ferriere R (2012) The fundamental role of competition in the ecology and evolution of mutualisms. In: Mousseau TA, Fox CW (eds) Year in evolutionary biology. Wiley, New York, pp 66–88

7. Shou WY, Ram S, Vilar JMG (2007) Synthetic cooperation in engineered yeast populations. Proc Natl Acad Sci U S A 104:1877–1882

8. Harcombe W (2010) Novel cooperation experimentally evolved between species. Evolution 64:2166–2172

9. Hosoda K, Suzuki S, Yamauchi Y, Shiroguchi Y, Kashiwagi A, Ono N et al (2011) Cooperative adaptation to establishment of a synthetic bacterial mutualism. PLoS ONE 6:e17105

10. Summers ZM, Fogarty HE, Leang C, Franks AE, Malvankar NS, Lovley DR (2010) Direct exchange of electrons within aggregates of an evolved syntrophic coculture of anaerobic bacteria. Science 330:1413–1415

11. Hillesland KL, Stahl DA (2010) Rapid evolution of stability and productivity at the origin of a microbial mutualism. Proc Natl Acad Sci U S A 107:2124–2129

12. Stolyar S, Van Dien S, Hillesland KL, Pinel N, Lie TJ, Leigh JA et al (2007) Metabolic modeling of a mutualistic microbial community. Mol Syst Biol 3

13. Schink B, Stams AJ (2002) Syntrophism among prokaryotes. The prokaryotes: an evolving electronic resource for the microbiological community, 3rd edn, Release 3.8. http://link.springer-ny.com/link/service/books/10125/

14. Stams AJ, Plugge CM (2009) Electron transfer in syntrophic communities of anaerobic bacteria and archaea. Nat Rev Microbiol 7:568–577

15. Sieber JR, McInerney MJ, Gunsalus RP (2012) Genomic insights into syntrophy: the paradigm

for anaerobic metabolic cooperation. Annu Rev Microbiol 66:429–452

16. Bryant M, Wolin E, Wolin M, Wolfe R (1967) *Methanobacillus omelianskii*, a symbiotic association of two species of bacteria. Arch Mikrobiol 59:20–31

17. Zhang W, Culley DE, Scholten JC, Hogan M, Vitiritti L, Brockman FJ (2006) Global transcriptomic analysis of *Desulfovibrio vulgaris* on different electron donors. Antonie Van Leeuwenhoek 89:221–237

18. Scholten JCM, Conrad R (2000) Energetics of syntrophic propionate oxidation in defined batch and chemostat cocultures. Appl Environ Microbiol 66:2934–2942

19. Scholten JC, Culley DE, Brockman FJ, Wu G, Zhang WW (2007) Evolution of the syntrophic interaction between *Desulfovibrio vulgaris* and *Methanosarcina barkeri*: involvement of an ancient horizontal gene transfer. Biochem Biophys Res Commun 352:48–54

20. Walker CB, He ZL, Yang ZK, Ringbauer JA, He Q, Zhou JH et al (2009) The electron transfer system of syntrophically grown *Desulfovibrio vulgaris*. J Bacteriol 191:5793–5801

21. Walker CB, Redding-Johanson AM, Baidoo EE, Rajeev L, He Z, Hendrickson EL et al (2012) Functional responses of methanogenic archaea to syntrophic growth. ISME J 6(11):2045–2055

22. Zhou J, He Q, Hemme CL, Mukhopadhyay A, Hillesland K, Zhou A et al (2011) How sulphate-reducing microorganisms cope with stress: lessons from systems biology. Nat Rev Microbiol 9:452–466

23. Hungate R (1969) A roll tube method for cultivation of strict anaerobes. Meth Microbiol 3B:117–132

24. Plugge CM (2005) Anoxic media design, preparation, and considerations. Meth Enzymol 397:3–16

25. Widdel F, Bak F (1992) Gram-negative mesophilic sulfate-reducing bacteria. In: Balows A et al (eds) The prokaryotes: a handbook on the biology of bacteria: ecophysiology, isolation, identification, applications. Springer, New York, pp 3352–3378

26. Whitman WB, Shieh J, Sohn S, Caras DS, Premachandran U (1986) Isolation and characterization of 22 mesophilic *Methanococci*. Syst Appl Microbiol 7:235–240

27. Mukhopadhyay A, Redding AM, Joachimiak MP, Arkin AP, Borglin SE, Dehal PS et al (2007) Cell-wide responses to low-oxygen exposure in *Desulfovibrio vulgaris* Hildenborough. J Bacteriol 189:5996–6010

28. Jarrell KF (1985) Extreme oxygen sensitivity in methanogenic archaebacteria. Bioscience 35:298–302

29. Jones WJ, Paynter MJB, Gupta R (1983) Characterization of *Methanococcus maripaludis* sp nov, a new methanogen isolated from salt marsh sediment. Arch Microbiol 135:91–97

30. Postgate JR, Campbell LL (1966) Classification of *Desulfovibrio* species nonsporulating sulfate-reducing bacteria. Bacteriol Rev 30:732–738

31. Kuever J, Rainey F, Widdel F (2005) Family I. *Desulfovibrionaceae fam.* nov. In: Garrity G et al (eds) Bergey's manual of systematic bacteriology. Springer, New York, pp 926–938

Chapter 9

Therapeutic Microbes for Infectious Disease

Choon Kit Wong, Mui Hua Tan, Bahareh Haji Rasouliha, In Young Hwang, Hua Ling, Chueh Loo Poh, and Matthew Wook Chang

Abstract

The rapid emergence of multidrug-resistant pathogens has invoked concerns of our current limitations in controlling the spread of infectious disease. To resolve this, we have applied synthetic biology principles to engineer human commensal microbe that can specifically sense and kill an antibiotic-resistant strain of *P. aeruginosa*. In this chapter, we describe the methods used to assemble, characterize, and evaluate the effectiveness of our engineered microbe in multicellular systems.

Key words Synthetic biology, Genetic circuits, Therapeutic microbes, Pyocin, Quorum sensing

1 Introduction

The effectiveness of antibiotic treatment against human pathogens is increasingly compromised by the rapid emergence of multidrug-resistant superbugs. Certainly, our current strategies against infectious diseases are limited and unsustainable in the long term, and novel approaches that directly address the challenges posed by deadly pathogens must be developed. Among multidrug-resistant pathogens, *Pseudomonas aeruginosa*, a leading cause of hospital-acquired infection, is known to be a major risk factor in the recovery of immunocompromised patients such as those suffering from cancers and surgical transplants and the aging population [1, 2]. With *P. aeruginosa* infection increasingly recognized as an enteric disease model [3], we have sought to apply synthetic biology principles in engineering a strain of *Escherichia coli* that can sense *P. aeruginosa* and subsequently activate downstream mechanisms leading to the production and release of killing molecules against the pathogen [4]. To implement our concept, we genetically functionalized *E. coli* with a quorum sensing (QS) device derived from *P. aeruginosa* for the detection of homoserine lactones secreted by the pathogen. On activation, the engineered *E. coli* will synthesize and accumulate pyocin S5, a pore-forming bacteriocin which is

Lianhong Sun and Wenying Shou (eds.), *Engineering and Analyzing Multicellular Systems: Methods and Protocols*, Methods in Molecular Biology, vol. 1151, DOI 10.1007/978-1-4939-0554-6_9, © Springer Science+Business Media New York 2014

117

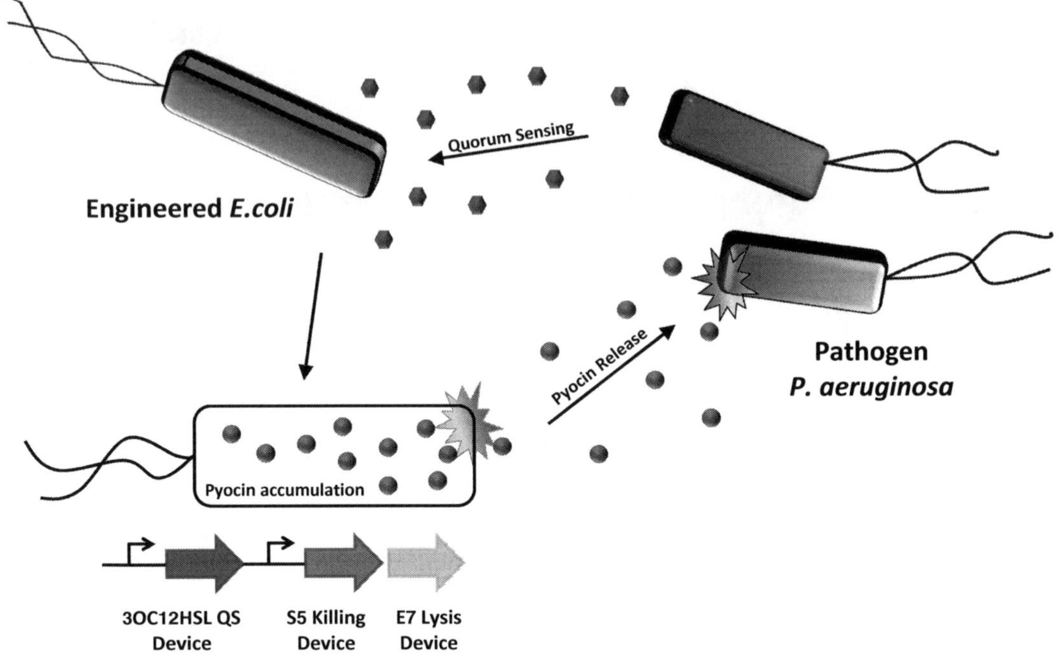

Fig. 1 Schematic of sense–kill system. Engineered *E. coli* carrying the QS-S5-E7 construct expresses pyocin S5 and E7 lysis proteins upon detection of diffusible quorum sensing molecules secreted by *P. aeruginosa*. Thereafter, pores generated on cell surface of the engineered *E. coli* facilitate the release of accumulated pyocin molecules, leading to effective killing of the pathogen

specific against our target pathogen, before initiating lysis to release the accumulated killing molecules against the pathogen (Fig. 1). Herein, we present the framework and procedures for the characterization and preliminary assessment of an engineered microbe that can be applied in treatment against infectious disease. This includes system assembly, characterization of quorum sensing device, scanning electron microscopy analysis of *E. coli* cytoplasmic release, and in vitro antimicrobial assays in multicellular systems.

2   Materials

2.1 System Assembly

1. Thermocycler.
2. Gel electrophoresis setup.
3. UV transilluminator.
4. Heating block.
5. Nanodrop (Thermo Scientific).
6. Electroporator.
7. Isothermal water bath.
8. Shaking incubator.

Table 1
Preparation of 5× isothermal buffer (ISO)

Initial concentration	Volume in 6 ml	Final concentration	Source
Sterile ddH$_2$O	Top up to 6 ml		
Tris–HCl pH 7.5 (1 M)	3 ml	500 mM	
MgCl$_2$ (2 M)	150 µl	50 mM	
dNTP mix (10 mM of ea. dNTP)	600 µl	1 mM of ea. dNTP	
DTT (1 M)	300 µl	50 mM	Thermo Scientific
NAD (100 mM)	300 µl	5 mM	Applichem Lifescience
PEG-8000	1.5 g	25 % w/v	Affymetrix

9. Sterile scalpel.

10. Microcentrifuge tubes 1.5 ml.

11. PCR reaction tubes.

12. Electroporation cuvette 2.0 mm.

13. L-shape cell spreader.

14. Round-bottom cell culture tubes with cap.

15. Phusion High Fidelity DNA polymerase (New England Biolabs), 5× Phusion HF buffer, 100 % DMSO, dNTPs.

16. 1× TAE buffer: 40 mM Tris base, 40 mM acetic acid, 1 mM EDTA. Prepared from 10× Tris–acetate–EDTA buffer by diluting with sterile deionized water to 1 l.

17. 0.8 % agarose gel with eight wells in 55 ml 1× TAE and 1× nucleic acid gel stain. This is sufficient for loading of up to 50 µl of PCR product.

18. QIAquick Gel Extraction Kit (Qiagen).

19. Gibson Assembly Master Mix (New England Biolabs): Alternatively, a 5× isothermal reaction buffer (ISO) and 1.33× assembly master mix can be prepared with the recipe formulated by Gibson as described in Tables 1 and 2 [5]. The 5× ISO buffer and 1.33× assembly master mix can be stored at –20 °C in aliquots of 350 and 15 µl for up to a year, respectively.

20. SOC medium: 0.5 % yeast extract, 2 % tryptone, 10 mM NaCl, 2.5 mM KCl, 10 mM MgCl$_2$, 10 m M MgSO$_4$, 20 mM glucose. Glucose is added last after autoclaving the other ingredients. Sterilize the solution with a 0.20 µm filter and store at 4 °C.

21. Chemically competent *E. coli* cells.

22. Antibiotic-supplemented LB agar plate.

Table 2
Preparation of 1.33× assembly master mix

Initial concentration	Volume in 1.2 ml	Final concentration	Source
Sterile ddH_2O	Top up to 1.2 ml		
ISO buffer (5×)	320 µl	500 mM	
T5 exonuclease (10 U/µl)	0.64 µl	6.4 U	New England Biolabs
Phusion polymerase (2 U/µl)	20 µl	40 U	New England Biolabs
Taq ligase (40 U/µl)	160 µl	6,400 U	New England Biolabs

Table 3
Preparation of characterization media (M9S)

Initial concentration	Volume in 1 l	Final concentration
5× M9 minima salts	5.64 g	1×
Thiamine hydrochloride	150 mg	1 mM
$MgSO_4$ (0.1 M)	20 ml	2 mM
$CaCl_2$ (0.5 M)	200 µl	0.1 mM
Casein hydrolysate (10 % w/v)	20 ml	0.2 %
Glycerol (50 % v/v)	8 ml	0.4 %
Sterile ddH_2O	Top up to 1 l	

2.2 Characterization of Quorum Sensing Device

1. Shaking incubator.
2. Microplate reader with programmable time-based fluorescence and absorbance measurement functions (Biotek, Synergy HT Multi-Mode Microplate Reader recommended).
3. Gen5 Data Analysis software (Biotek, software for Synergy HT microplate reader).
4. Matlab with Curve Fitting Toolbox (Mathworks, Natick, MA, USA).
5. MS Excel (Microsoft Office 2010 or equivalent).
6. Incubator.
7. Multichannel pipette.
8. Vortex.
9. Microplate sealing film.
10. Round-bottom cell culture tubes with cap.
11. Petri dishes.
12. 1 l of supplemented M9 characterization media: 1× M9 salts, 1 mM thiamine hydrochloride, 0.4 % glycerol, 0.2 % casein hydrolysate, 2 mM $MgSO_4$, and 0.1 mM $CaCl_2$ (Table 3).

Sterilize the solution with a 0.20 μm filter and store at 4 °C away from light.

13. 3OC$_{12}$ homoserine lactone: Dissolved in absolute DMSO to a stock concentration of 10 mM. Obtain 3OC$_{12}$HSL test solutions (0.1 μM, 1 μM, 10 μM, 0.1 mM, and 1 mM) by sequential 10× dilution in supplemented M9 characterization medium (10 μl 3OC$_{12}$HSL solutions to 90 μl supplemented M9).

14. Fresh plate of *E. coli* that expresses green fluorescent protein (GFP) when induced with 3OC$_{12}$HSL (QS-GFP, Amp[r]).

2.3 Scanning Electron Microscopy Analysis of E. coli Cytoplasmic Release

1. Shaking incubator.

2. Centrifuge.

3. Microcentrifuge with refrigeration function.

4. Vacuum dryer.

5. Osmium plasma coater.

6. Field emission scanning electron microscope (FESEM).

7. Round-bottom cell culture tubes with cap.

8. Microcentrifuge tubes 2.0 ml.

9. Membrane filters 0.2 μm.

10. Syringes 5 ml.

11. 24-well microplate plate.

12. Tweezer.

13. 10 ml of LB broth with ampicillin (100 μg/ml).

14. PEI-coated silicon slide: To prepare a silicon substrate for SEM application, first clean the surface of a silicon slide with 70 % ethanol solution and blow free of dust with an air hose. Immerse the slide in polyethyleneimine (PEI) for 30 min before rinsing completely with sterile deionized water 2–3 times. Leave to dry on a clean bench for at least 3 h or until use. PEI-coated slides can be stored in for up to 2 weeks at room temperature.

15. 0.1 M sodium cacodylate, pH 7.4.

16. 2.5 % w/v glutaraldehyde in 0.1 M sodium cacodylate, pH 7.4.

17. 1 % w/v osmium tetroxide in 0.1 M sodium cacodylate, pH 7.4.

18. Ethanol solutions (37, 67, 95, and 100 %), 5 ml of each.

19. SEM coating powder, gold–palladium alloy (60:40).

20. Fresh plate of engineered *E. coli* with sense–kill–release functions (QS-S5-E7, Amp[r]).

21. Control plate of engineered *E. coli* with only sense–kill functions (QS-S5, Amp[r]).

2.4 Overlay Inhibition Assay

1. Kitchen microwave.
2. Incubator with shaking function.
3. Isothermal water bath.
4. Bio-imager (Biorad ChemiDoc XRS or equivalent).
5. Round-bottom cell culture tubes with cap.
6. Microcentrifuge tubes 2.0 ml.
7. Membrane filters 0.2 μm.
8. Syringes 5 ml.
9. Tryptic soy agar plates (3 % w/v Bacto tryptic soy broth and 1.5 % w/v Bacto agar in 400 ml of deionized water): Autoclave the mixture, and transfer 12 ml aliquots of the resultant solution into sterile petri dishes. Solidified tryptic soy agar (TSA) plates can be stored at 4 °C for up to a month.
10. 100 ml of soft agar (1 % peptone w/v and 0.5 % w/v Bacto agar in 100 ml deionized water): Autoclave the mixture and cool to 45 °C in a water bath for immediate use. Otherwise, the solution can be stored at room temperature for up to a month.
11. Ice bath.
12. Fresh plate of *P. aeruginosa* clinical isolate In7 which produces $3OC_{12}HSL$.
13. Fresh plate of engineered *E. coli* with sense–kill–release functions (QS-S5-E7, Amp^r).
14. Control plate of engineered *E. coli* with only sense–kill functions (QS-S5, Amp^r).

2.5 Co-culture Inhibition Assay of Engineered E. coli and P. aeruginosa

1. Microplate reader with programmable time-based fluorescence functions (Biotek, Synergy HT Multi-Mode Microplate Reader recommended).
2. MS Excel (Microsoft Office 2010 or equivalent).
3. 96-well microplate (Greiner 96-well CellStar®, black).
4. Round-bottom cell culture tubes with caps.
5. 10 ml of LB broth with ampicillin (100 μg/ml).
6. Optional: LB agar plate with chloramphenicol (100 μg/ml).
7. Fresh plate of *P. aeruginosa* clinical isolate In7 carrying plasmid pMC-PA$_{GFP/CM}$ (*see* **Note 1**).
8. Fresh plate of engineered *E. coli* with sense–kill–release functions (QS-S5-E7, Amp^r).

3 Methods

3.1 System Assembly

Gibson assembly is an isothermal "multi-pot" cloning technique that harnesses the collective enzymatic actions of T5 exonuclease, Phusion DNA polymerase, and Taq DNA ligase to assemble

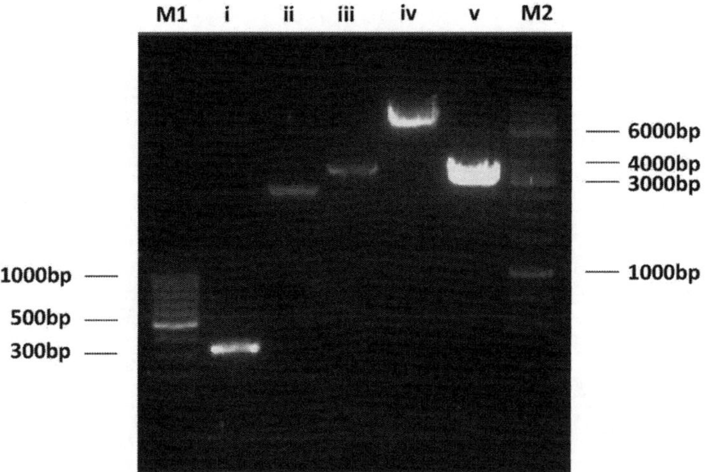

Fig. 2 Gibson assembly of sense–kill system QS-S5-E7. The image shows gel electrophoretic analysis of the final assembled construct and the associated PCR-verified modules: (*i*) E7 insert ~320 bp, (*ii*) QS-S5 insert ~2.8 kb, (*iii*) linearized vector ~3.5 kb, (*iv*) linearized QS-S5-E7-vector ~6.6 kb, and (*v*) double-digested fragments of QS-S5-E7 ~3.1 kb and linearized vector ~3.5 kb. The molecular weight markers used in this figure were (*M1*) NEB 100 bp and (*M2*) NEB 1 kb DNA ladder. Electrophoretic separation was performed with approximately 300 ng of DNA on a 0.8 % agarose gel at 120 V for 30 min

genetic constructs of up to several hundred kilobases. Originally developed for the assembly of synthetic genomes, it is now widely used for fast and efficient pathway reconstruction [5]. In this demonstration, we constructed our final system (QS-S5-E7) using a customized protocol of the Gibson assembly technique (Fig. 2).

1. Perform an in silico design of the overall genetic construct on plasmid drawing software. Zoom into the sequence region 50 bp upstream and downstream on the plasmid vector which also contains the desired genetic construct.

2. Design and synthesize primers for PCR amplification of linear inserts and vector, aiming for target fragment sizes of 300 bp–5 kb (*see* **Note 2**). Primers should be designed to incorporate between 20 and 40 bp of overlapping sequences, adding spacer sequences where necessary (*see* **Notes 3** and **4**).

3. Set up PCR reaction tubes on ice as described in Table 4. We recommend the use of Phusion High-Fidelity DNA polymerase for the amplification of long DNA fragments.

4. Run PCR reactions in a thermocycler with cycling conditions as described in Table 5. Of note, successful PCR products are generated using a two-cycle PCR approach. During the first cycle, DNA template is enriched for five cycles by using the T_m which overlapped with the source DNA template to determine the annealing temperature. The second cycle adopts a higher

Table 4
PCR reaction composition

Initial concentration	Volume per 40 µl reaction	Final amount in 40 µl
Sterile ddH$_2$O	Up to 40 µl	
Phusion PCR buffer (5×)	8 µl	1×
DMSO (100 %)	2 µl	5×
dNTP mix (40 mM)	0.8 µl	0.8 mM
Forward primer	0.8 µl	0.5 µM
Reverse primer	0.8 µl	0.5 µM
DNA template (1–5 ng)	Variable	5–20 ng
Phusion high-fidelity DNA polymerase (2 U/µl)	0.4 µl	1 U

Table 5
PCR cycling conditions

Cycle no.	Denaturation	Annealing	Extension
1	98 °C, 30 s		
2–6 (first cycle)	98 °C, 10 s	T_{m1} + 3 °C, 30 s	72 °C, 15 s/kb
7–26 (second cycle)	98 °C, 10 s	T_{m2} + 3 °C, 30 s	72 °C, 15 s/kb
27			72 °C, 5× of extension

T_m of the enriched template for more specific priming of target amplicons. Here, we demonstrate the amplification of three DNA fragments using source templates from our plasmid repository, with sizes of 2.8 kb (QS-S5), 320 bp (E7), and 3.5 kb (vector).

5. Isolate the amplified DNA fragments on a 0.8 % agarose gel by gel electrophoresis (85 V, 45 min).

6. Visualize the agarose gel on a UV transilluminator, and excise the band of interest with a sterile scalpel. Purify the isolated DNA fragments with QIAquick Gel Extraction Kit according to the vendor's instruction, and analyze each purified fragment with Nanodrop. Of note, good DNA concentrations above 20 ng/µl are ideal for efficient Gibson assembly reactions (*see* **Note 5**). Hold the purified DNA fragments on ice until use.

7. Thaw a 15 µl aliquot of 1.33× assembly master mix for 5 min on ice.

8. Add equimolar amount of DNA inserts and vector (total DNA concentration of up to 100 ng) to 15 µl of assembly master mix.

Top up with nuclease-free water to a total reaction volume of 20 µl. In this example, our system was successfully assembled with 20 ng of linearized vector DNA. The amount of insert required can be empirically calculated by Eq. 1:

$$\text{Mass of insert}\,(\text{ng}) = \frac{\text{Length of insert}\,(\text{bp})}{\text{Length of vector}\,(\text{bp})} \times \text{Mass of vector}\,(\text{ng}). \tag{1}$$

9. Incubate the reaction mixture at 50 °C in a thermocycler with heated lid for 1 h. The temperature inversion is necessary to avoid evaporation and the consequential loss of enzymatic activity. On completion, hold the reaction tube on ice for at least 15 min.

10. Thaw 50 µl aliquot of chemically competent *E. coli* cells for 5 min on ice. Transform 5 µl of reaction mixture into *E. coli* cells by 45 s heat shock at 42 °C and 5-min incubation on ice. Add 400 µl of prewarmed SOC, and transfer the entire transformation culture to a sterile round-bottom capped tube. Incubate the culture in a shaking incubator set at 225 rpm and 37 °C for 40 min.

11. Spread 1/5 of the transformation culture (~100 µl) on antibiotic-selective plate. Resuspend the remaining 4/5 of the culture in 100 µl of SOC and plate on another antibiotic-selective plate.

12. Incubate the plates in a 37 °C incubator for 15 h, and characterize positive clones the next day by colony PCR.

3.2 Characterization of Quorum Sensing Device

To evaluate the sensitivity and range of our quorum sensing device in an *E. coli* host, we cloned a GFP reporter downstream of the sensing device (QS-GFP) and measured the fluorescence output after induction with synthetic $3OC_{12}HSL$. Consequently, the characterization profile of this biosensor can be used to estimate the amount of $3OC_{12}HSL$ that is secreted by *P. aeruginosa* clinical isolates. M9 medium is supplemented with appropriate antibiotic throughout this experiment unless otherwise stated.

1. Inoculate a single colony of *E. coli* with the plasmid QS-GFP from a fresh plate into 5 ml of M9 characterization medium. Incubate the culture in a 37 °C shaking incubator set at 225 rpm for 15 h.

2. After 15 h of incubation, dilute the culture 100× in 10 ml of prewarmed M9 characterization medium. Allow the re-inoculated culture to grow to OD_{600} of 1.0 under the same conditions. This corresponds to mid-log phase of the cells.

3. While waiting for the re-inoculated culture to grow to the desired cell density, prefill a 96-well microplate with 110 µl of fresh M9 characterization medium. Vortex $3OC_{12}HSL$ test

solutions (0.1 μM, 1 μM, 10 μM, 0.1 mM, and 1 mM) evenly, and add 1–2 μl of each $3OC_{12}HSL$ solution in triplicates so that the final $3OC_{12}HSL$ concentration in the 96-well microplate is in the range of $0–10^{-5}$ M.

4. Fill any empty wells with 220 μl of M9 characterization medium to serve as both sterility control and liquid reservoir. Ensure that at least a column of wells is allocated for zeroing with 220 μl of M9 characterization medium. Prewarm the microplate in a 37 °C incubator for 30 min just before use; otherwise, store at 4 °C. Ensure that the microplate is covered with lid and sealed tight with parafilm when outside a biological hood or Bunsen flame.

5. Prepare the Synergy HT Multi-Mode Microplate Reader for endpoint measurement of absorbance (OD_{600}) and fluorescence (485 nm excitation/528 nm emission) at regular intervals of 5 min with 30 s of rapid shaking before sampling. Temperature is set to 37 °C. Advance users can further program Biotek's Gen5 Data Analysis software to obtain triplicate average of background-subtracted fluorescence per OD_{600}. Otherwise, data analysis can be performed in an exported Excel sheet.

6. On reaching the desired cell density, pour the re-inoculated culture into a sterile petri dish. Transfer 110 μl of re-inoculated culture into the prefilled microplate with a multichannel pipette, effectively diluting the culture to OD 0.5. Mix the culture evenly by pipetting, and seal the microplate with a transparent sealing film. Note that the culture is induced at this point.

7. Transfer the microplate into the microplate reader, and begin measurement. Run the equipment for 3 h.

8. Export the raw experimental data into MS Excel. Normalize the fluorescence (RFU) and absorbance (OD_{600}) values of each well by zeroing with the average of pure M9 characterization medium as shown in Eqs. 2 and 3. The relative quantity of GFP molecules in a single cell is derived as a ratio of background-subtracted fluorescence to OD_{600} values (RFU/OD_{600}):

$$\text{Normalized fluorescence of sample } x \text{ at time } t, \text{RFU}_t^{x'} = \text{RFU}_t^x - \text{RFU}_t^{M9}. \qquad (2)$$

$$\text{Normalized OD600 of sample } x \text{ at time } t, \text{OD}_t^{x'} = \text{OD}_t^x - \text{OD}_t^{M9}. \qquad (3)$$

9. Obtain an average of RFU/OD_{600} from the technical triplicates. Repeat for a total of at least three independent experiments. Use only analyzed data from independent experiments (biological replicates) to determine the standard deviations of statistical means.

10. Obtain the GFP production rate ($RFU.OD_{600}^{-1}/min$) by taking the difference of RFU/OD_{600} values from two time points and dividing the result by the time interval δt as shown in Eq. 4:

GFP production rate of sample x at time t,

$$\frac{\mathrm{RFU}}{\mathrm{OD} \cdot t} = \frac{\mathrm{RFU}^{x}_{t+\delta t} / \mathrm{OD}^{x}_{t+\delta t} - \mathrm{RFU}^{x}_{t} / \mathrm{OD}^{x}_{t}}{\delta t}. \tag{4}$$

11. Plot the analyzed data on a 3D surf plot with Matlab, setting $3OC_{12}HSL$ concentration, time, and GFP production rate as the x–y–z-axis, respectively.

12. Identify the period to which steady-state GFP production rate is observed with the 3D graphical plot obtained in **step 10**. Then, determine the average GFP production rate by averaging the numerical values of all GFP production rates that fall within this identified period (*see* **Note 6**). Fit the experimental results to an empirical mathematical model as shown in Eq. 5 using Matlab Curve Fitting Toolbox where A, B, C, and n are curve-fitted empirical parameters:

$$\text{GFP production rate, } \Upsilon = A + \frac{B\left[3OC_{12}HSL\right]^{n}}{C^{n} + \left[3OC_{12}HSL\right]^{n}}. \tag{5}$$

3.3 Scanning Electron Microscopy Analysis of E. coli Cytoplasmic Release

This protocol describes a method to visually inspect the extent of lysis and cytoplasmic release of *E. coli* with scanning electron microscopy. The fixation method discussed here can also be used for surface morphology studies of other bacteria species.

1. Inoculate single colonies of engineered *E. coli* (QS-S5-E7) and control *E. coli* (QS-S5) into cell culture tubes each with 10 ml of LB plus ampicillin. Incubate the cell cultures in a 37 °C shaking incubator set at 225 rpm for 15 h.

2. Dilute each culture to $OD_{600} \sim 0.01$ in 10 ml of LB plus ampicillin. Allow growth to a final cell density of $OD_{600} \sim 0.5$ in the same culture conditions as described above.

3. Optional step: Further dilute *E. coli* cultures 10× in 10 ml of LB plus ampicillin. Allow growth to a final cell density of $OD_{600} \sim 0.5$ (*see* **Note 7**).

3. Induce both the engineered and control *E. coli* by adding 2 μl of 0.1 mM $3OC_{12}HSL$ to 2 ml of each culture. Incubate the induced *E. coli* cultures in a 37 °C shaking incubator set at 225 rpm for 2 h. Perform the experiment in duplicates.

4. After 2 h of induction, determine the optical density of each culture. The optical density of control *E. coli* culture should be significantly larger than the engineered *E. coli*. Transfer each sample to sterile 2 ml microcentrifuge tubes. Centrifuge at $1,500 \times g$ and 4 °C for 10 min, and discard the supernatant.

5. Wash the cells of each sample with 1 ml of 0.1 M sodium cacodylate and resuspend gently by pipetting. Centrifuge at $1,500 \times g$ and 4 °C for 10 min, and carefully discard the supernatant. Repeat the washing procedure for a total of three washes.

6. Perform primary fixation by gently resuspending the cell pellet obtained from **step 7** with 1 ml of 2.5 % w/v glutaraldehyde in 0.1 M sodium cacodylate. Incubate at 4 °C for 2 h or overnight.

7. Repeat the washing procedures in **step 7** with 1 ml of 0.1 M sodium cacodylate thrice.

8. Gently resuspend the cell pellet obtained from **step 9** with a small volume of 0.1 M sodium cacodylate (10–50 µl) (*see* **Note 8**).

9. Place a PEI-coated silicon slide on a sterile petri dish with tweezers.

10. Spot 2 µl of engineered cell culture (QS-S5-E7) from **step 10** on a predefined corner of a PEI-coated silicon slide. Diagonally across the same slide, spot 2 µl of control cell culture without the lysis device (QS-S5). Cover the petri dish with its lid, and incubate the silicon slide at 25 °C for 30 min.

11. Optional step: Transfer the loaded slide with a tweezer to a well on 24-well microplate that contains 1 ml of 1 % w/v osmium tetroxide in 0.1 M sodium cacodylate. Ensure that the loaded slide is totally immersed for secondary fixation and incubate at 25 °C for 90 min (*see* **Note 9**).

12. Fill six other wells on the 24-well microplate with 1 ml of ethanol solution (37, 67, 95, 100, 100, and 100 %). Carefully dehydrate the loaded silicon slide by immersing the slide in serial concentration of absolute ethanol. Allow dehydration at 25 °C for 15 min (*see* **Notes 10** and **11**).

13. Dry the slide in a vacuum dryer overnight at 25 °C for 15 min.

14. Coat the biological samples with 20 nm of gold–palladium alloy with an osmium plasma coater and examine using a FESEM at 10 kV.

3.4 Overlay Inhibition Assay

The overlay inhibition assay described here provides a method for the in vitro evaluation of antimicrobial efficacy of pyocin produced and released by an engineered *E. coli* (Fig. 3). The protocol may be modified accordingly to identify the minimum inhibitory concentration (MIC) of other antimicrobial compounds [6].

1. Inoculate single colonies of *P. aeruginosa* isolate In7, engineered *E. coli* (QS-S5-E7), and control *E. coli* (QS-S5) into cell culture tubes each with 10 ml of LB plus ampicillin. Incubate the cell cultures in a 37 °C shaking incubator set at 225 rpm for 15 h.

2. Dilute each culture to $OD_{600} \sim 0.1$ in 10 ml of LB plus ampicillin. Allow growth to a final cell density of $OD_{600} \sim 1.0$ in the same culture conditions as described above.

3. Separate 10 ml of In7 culture into 2 and 8 ml, respectively. Hold both re-inoculated *E. coli* cultures (QS-S5-E7 and

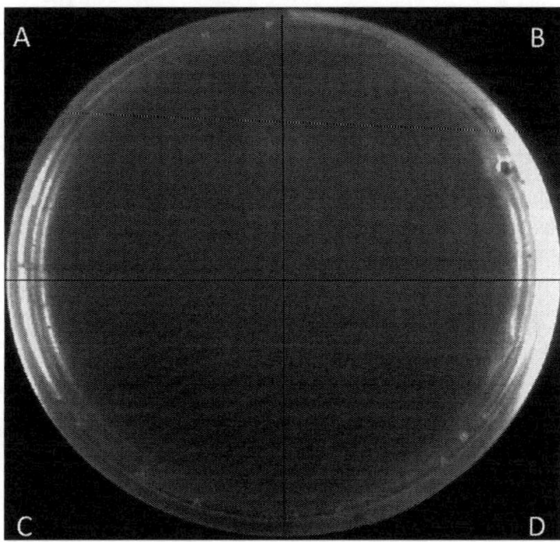

Fig. 3 Overlay inhibition assay of *P. aeruginosa*. Agar overlay of *P. aeruginosa* is spotted with the filtered supernatant of (**a**) QS-S5-E7 induced with 10–7 M AHL, (**b**) QS-S5-E7 induced with *P. aeruginosa* In7 supernatant, (**c**) uninduced engineered *E. coli* with full system (QS-S5-E7), and (**d**) QS-S5 induced with *P. aeruginosa* In7 supernatant

QS-S5) and 2 ml of In7 culture on ice until use. This step stops growth and preserves a constant cell density temporarily.

4. Centrifuge 8 ml of the re-inoculated In7 culture at $4{,}500 \times g$ for 10 min. Sterilize the supernatant with a 0.2 μm membrane filter, and transfer the filtered supernatant into a labelled round-bottom capped tube. Hold on ice until use. Otherwise, the filtered supernatant containing $3OC_{12}HSL$ from *P. aeruginosa* can be stored at –20 °C for up to a week.

5. Centrifuge 4 ml cultures of both the engineered and control *E. coli* at $4{,}500 \times g$ for 10 min, and discard the supernatant. Induced both the engineered and control *E. coli* by adding equal volume of filtered In7 supernatant as the volume of *E. coli* cultures which are discarded in each tube. Incubate the induced *E. coli* cultures in a 37 °C shaking incubator set at 225 rpm for 5 h.

6. Centrifuge 1 ml of each induced *E. coli* cultures at $4{,}500 \times g$ for 10 min. Sterilize the supernatant of induced *E. coli* cultures with 0.2 μm membrane filters, and transfer the filtered supernatants into labelled round-bottom capped tubes. Hold on ice until use.

7. Prewarm a TSA plate in a 37 °C incubator for 15 min.

8. Melt soft agar in a microwave for 2 min. Cool the bottle of soft agar under running tap water while maintaining the agar in a liquid state. Transfer 2.7 ml of soft agar to a round-bottom capped tube and hold in a water bath set at 55 °C.

Table 6
Mixing ratio of *P. aeruginosa* and engineered *E. coli* in co-culture inhibition assay

P. aeruginosa (ml)	*E. coli* (ml)	LB plus ampicillin (ml)	EC/PA ratio
0.2	0	1.8	0
0.2	0.2	1.6	1
0.2	0.4	1.4	2
0.2	0.6	1.2	3
0.2	0.8	1	4

9. Add 100 μl of In7 culture at $OD_{600} \sim 0.2$–2.4 ml of soft agar, and mix uniformly by vigorous shaking.

10. Transfer 2.5 ml of In7 soft agar culture onto a prewarmed TSA plate and spread uniformly by tilting the plate. Ensure that the agar plate is left on a horizontal surface with a water leveller. Cover the plate with lid, and allow the agar overlay to solidify at room temperature for 1 h.

11. Spot 10 μl of filtered supernatants from induced engineered *E. coli* and control *E. coli* cultures onto the In7 agar overlay and leave to dry in a biological hood for 1 h.

12. Incubate the plate in a 37 °C incubator for 6 h or more. Analyze the plate with ChemiDoc XRS bio-imager on the epiwhite mode or with a high-resolution digital camera.

3.5 Co-culture Inhibition Assay of Engineered E. coli and P. aeruginosa

The co-culture inhibition assay discussed here provides a method to determine the ratio of engineered *E. coli* (EC) to *P. aeruginosa* (PA) cells that is required to completely inhibit the proliferation of *P. aeruginosa*. The protocol may be modified accordingly for the co-culture of other cell lines.

1. Inoculate single colonies of *P. aeruginosa* isolate In7 (with plasmid pMC-PA$_{GFP/CM}$) and engineered *E. coli* (QS-S5-E7) into cell culture tubes, each with 10 ml of LB plus ampicillin. Incubate the cell cultures in a 37 °C shaking incubator set at 225 rpm for 15 h.

2. Dilute each culture to $OD_{600} \sim 0.1$ in 10 ml of LB plus ampicillin. Allow growth to a final cell density of $OD_{600} \sim 1.0$ in the same culture conditions as described above.

3. Mix In7 and engineered *E. coli* cultures in round-bottom capped tubes as described in Table 6 on ice (*see* **Note 12**). Top up to a total volume of 2 ml with LB plus ampicillin.

4. Incubate the mixed cultures in a 37 °C shaking incubator set at 225 rpm for 12 h.

5. Prepare the Synergy HT Multi-Mode Microplate Reader for endpoint measurement of fluorescence (485 nm excitation/ 528 nm emission).

6. At regular intervals of 3 h, transfer 100 µl aliquot of the mixed cultures into a 96-well microplate for a total of two technical replicates. Then, transfer the 96-well microplate to the microplate reader and measure for background-subtracted green fluorescence with pure LB plus ampicillin as the zeroing medium.

7. Optional step: Serially dilute the mixed cultures in the 96-well plate 10× with pure LB plus ampicillin, and perform CFU count on a chloramphenicol-selective LB agar plate. Only *P. aeruginosa* cells which carry the chloramphenicol resistance marker will develop into viable colonies.

8. Top up the mixed cultures with 200 µl of fresh LB plus ampicillin so that the total co-culture volume is maintained at ~2.0 ml. Return the mixed cultures to the 37 °C shaking incubator with the same settings as described above.

9. Export the normalized results into MS Excel, and plot relative green fluorescence (in RFU) against time (h). The EC/PA ratio which is most effective in preventing the growth of *P. aeruginosa* is distinguished by a relatively constant GFP expression profile (*see* **Note 13**).

4 Notes

1. pMC-PA$_{\mathrm{GFP/CM}}$ carries both chloramphenicol and carbenicillin selection marker and drives constitutive expression of GFP in *P. aeruginosa* with a Lac promoter. It allows for quantification of *P. aeruginosa* viability by both fluorescence measurement and CFU count and can be used to image *Pseudomonas* biofilm by confocal fluorescence microscopy.

2. Inserts below 300 bp are recommended to be spliced together into longer fragments by overlap extension PCR. Splicing overlap extension (SOE) can be performed with equimolar ratio of DNA fragments, setting the largest fragment as the basis of comparison (100 ng). Ensure that the primers designed for SOE PCR have at least 20–25 bp of overlapping sequences and are free from secondary structures that may interfere with DNA recombination.

3. Primer design is a balance between economics and having sufficient overlapping sequence such that the T_{m} is greater than 50 °C. Most gene synthesis companies offer standard desalted primers at affordable rates for oligos of up to 50 bp.

4. Spacer sequences are incorporated when repeated gene sequences are present in the construct, such as the repetitive use of similar promoters and terminators. Essentially, spacer sequences should be designed with ~50 % GC content and distinctly different from other spacer sequences to prevent cross interference. Also, ensure that the spacer sequences (1) do not form hairpin structure in the 5′UTR or introduce unwanted transcription and translation termination (TAA, TGA, and TAG); (2) are free from possible Shine–Dalgarno sequences, start codons (ATG, GTG, and TTG), and the methylation site (GATC) of *E. coli*; and (3) are free from unwanted restriction sites.

5. The concentration of amplified DNA recovered with QIAquick Gel Extraction Kit can be enhanced by recovering the two 40 µl volumes of PCR reaction in a single column. The final elution volume with nuclease-free water can be reduced to 20 µl instead of 30 µl as stated by the kit's instructional manual. Alternatively, concentrate 30 µl of eluted DNA with a vacuum evaporator. Also, we have experienced higher DNA recovery during elution step by leaving the spin columns to stand in a 55 °C heat block for 2 min after nuclease-free water is added before collection.

6. In our previous work using pTet-LasR-pLux-GFP as the reporting device, we observed relatively uniform GFP production rate between 20 and 80 min after AHL induction [4]. Note that the period in which steady-state GFP production rate is observed can vary in disparate designs of quorum sensing devices. Thus, characterization on a case-by-case basis may be necessary.

7. This additional dilution step reduces the amount of aging cells with unhealthy surface morphology and enhances surface contrast between lysed and unlysed cells.

8. The precise volumes to resuspend the cell pellets depend on the amount of cells, and some trial-and-error optimization may be necessary. Generally after 2 h of induction, we observed that the engineered *E. coli* with lysis device could be resuspended in 10 µl while the control *E. coli* with much more viable cells could be resuspended in 40 µl for good comparison.

9. While primary fixation with glutaraldehyde serves to cross-linked proteins, secondary fixation with osmium tetroxide enables the fixation of lipids and is recommended for surface morphology studies. Note that osmium tetroxide is rapidly converted to osmium dioxide which is no longer able to fix cells on exposure to heat or light.

10. Prolonged dehydration may lead to cell shrinkage, especially at low ethanol concentration.

11. A secondary dehydration step with acetone/hexamethyl disilazane (HMDS) solvent can be performed to improve contrast by serial immersion of the loaded silicon slides in 2:1 acetone:HMDS, 1:2 acetone:HMDS, and 100 % HMDS (*see* ref. 7).

12. Here, we have established that an $OD_{600} \sim 1.0$ corresponds to 1.0×10^8 CFU/ml of *E. coli* cells with our spectrophotometer. Due to optical variability in different spectrophotometers, we advised that OD_{600} – CFU calibration curves be generated to determine the actual cell densities before mixing different cell cultures together.

13. GFP molecules are stable for more than 24 h, and therefore a constant level of GFP fluorescence still exists for growth-inhibited *P. aeruginosa* cells. Alternatively, a less stable variant of GFP reporter with shortened half-life may be used to assess directly the number of viable cells by flow cytometry.

Acknowledgements

This work was funded by the National Medical Research Council of Singapore (CBRG11nov109) and the National Research Foundation of Singapore (NRF-CRP5-2009-03). The first author wishes to dedicate this work to C.L.P., M.W.C., M.N., and S.T.

References

1. Chang W, Small DA, Toghrol F, Bentley WE (2005) Microarray analysis of *Pseudomonas aeruginosa* reveals induction of pyocin genes in response to hydrogen peroxide. BMC Genomics 6(1):115

2. Small DA, Chang W, Toghrol F, Bentley WE (2007) Comparative global transcription analysis of sodium hypochlorite, peracetic acid, and hydrogen peroxide on *Pseudomonas aeruginosa*. Appl Microbiol Biotechnol 76(5):1093–1105

3. Okuda J, Hayashi N, Okamoto M, Sawada S, Minagawa S, Yano Y, Gotoh N (2010) Translocation of *Pseudomonas aeruginosa* from the intestinal tract is mediated by the binding of ExoS to an Na, K-ATPase regulator, FXYD3. Infect Immun 78(11):4511–4522

4. Saeidi N, Wong CK, Lo T-M, Nguyen HX, Ling H, Leong SSJ, Poh CL, Chang MW (2011) Engineering microbes to sense and eradicate *Pseudomonas aeruginosa*, a human pathogen. Mol Syst Biol 7:521

5. Gibson DG, Young L, Chuang R-Y, Venter JC, Hutchison CA, Smith HO (2009) Enzymatic assembly of DNA molecules up to several hundred kilobases. Nat Methods 6(5):343–345

6. Ling H, Saeidi N, Rasouliha BH, Chang MW (2010) A predicted S-type pyocin shows a bactericidal activity against clinical *Pseudomonas aeruginosa* isolates through membrane damage. FEBS Lett 584(15):3354–3358

7. Kang A, Chang MW (2012) Identification and reconstitution of genetic regulatory networks for improved microbial tolerance to isooctane. Mol Biosyst 8(4):1350–1358

Part II

Analyzing and Modeling Multicellular Systems

Quantitative Measurement and Analysis in a Synthetic Pattern Formation Multicellular System

Xiongfei Fu and Wei Huang

Abstract

Pattern formation has been studied for more than a century in biology. In recent years there are increasing interests in studying it using bacteria and synthetic biology tools to program intercellular communication and cellular response to environment. Quantitative measurement is critical to dissect the interplay between the synthetic gene circuits with underline cellular processes and verify the mechanism determining the pattern formation. Here, we describe simple optical setups for quantitative measurements of the cell density and growth and spatial–temporal dynamic characterization of *E. coli* pattern formation in soft agar plates.

Key words Pattern formation, Synthetic biology, Quantitative biology, Quorum sensing, Time-lapse imaging, Optical density

1 Introduction

One of the fundamental questions in biology is how the cells coordinate their movement and fate to form well-organized spatial–temporal pattern [1–5]. It has been studied in the content of bacterial ecosystems, animal embryo development, as well as stem cell-based tissue engineering. Many molecular components and interactions have been well studied. However the underlying general principles are surprisingly difficult to be identified due to the overwhelming complex physiological context. For instance, the famous Turing model of lateral inhibition for pattern formation, one of the well-received theories by experimental biologists, has been published 60 years ago [6]. However no direct evidence can be found until recently when a group quantitatively measured the key differential diffusivity of Nodal and Lefty to confirm this model in zebra fish embryogenesis [7]. This highlights the importance and difficulty of quantitative biology measurement.

We took an alternative approach using synthetic biology tools [8] to specify and engineer the factors that control patterns formed by populations of *E. coli* cells. Specifically, we engineered the

Lianhong Sun and Wenying Shou (eds.), *Engineering and Analyzing Multicellular Systems: Methods and Protocols,*
Methods in Molecular Biology, vol. 1151, DOI 10.1007/978-1-4939-0554-6_10, © Springer Science+Business Media New York 2014

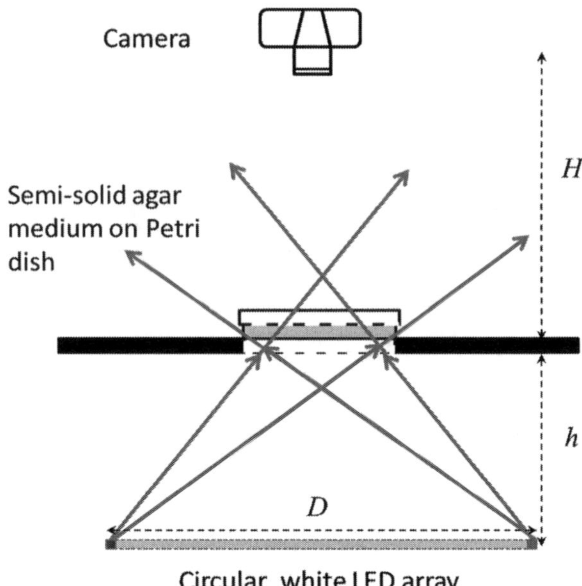

Fig. 1 A schematic view of the pattern formation imaging system. A circular white LED light belt illuminates the agar plate slantwise. Light was blocked except for right beneath the plate. A digital camera is positioned right over the plate so that LED light cannot directly reach it. The dimensions D, h, and H are chosen to enhance the contrast of the cell density variance

quorum sensing systems (LuxI) from *Vibrio fischeri* to *E. coli* cells so that the local cell density is represented by local concentration of signalling molecule N-acyl homoserine lactones (AHL). The same cells then sense the local cell density via LuxR-based gene regulation circuit to downregulate expression of CheZ gene and suppress cell motility. This control resulted in a sequential and periodic stripe formation. Careful quantitative measurement and modelling revealed the principle behind controlled individual cell behavior and pattern formation at the macroscale, which is then confirmed by experimental verification of model predicted pattern tuning method [9]. In this chapter we describe the optical methods to measure the spatial–temporal dynamics of the pattern and quantitative measurement of cell density profiles and growth rate in Petri dish that can be adapted for similar studies.

In our experiments, the patterns are formed in 0.15–0.35 % and 2 mm thick semisolid agar in Petri dish. The cell density has a typical optical density range of 0.1–3. So it is mostly transparent and very hard to view the patterns of subtle cell density variations. We have developed two easy methods to make instrumentations to measure the patterns. The first one is based on a light scattering principle that enhances the density contrast (Fig. 1). It enables fast two-dimensional pattern acquisition and high spatial resolution but does not provide quantitative density measurement. The second is

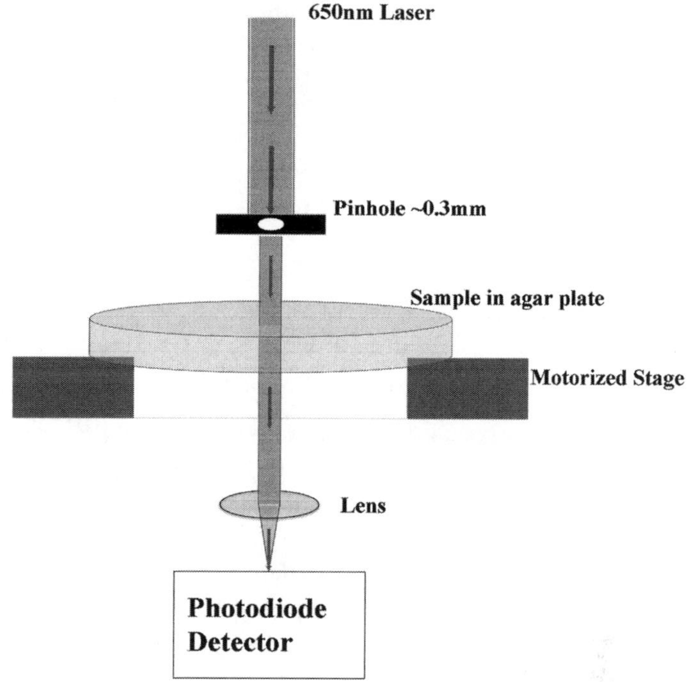

Fig. 2 A schematic view of the customized optical setup for the real-time measurement of the spatiotemporal cell density profile in semisolid agar dishes. A parallel 650-nm laser beam is guided through a 300-μm pinhole, passed through the sample in semisolid agar on the Petri dish at right angle, and collected via a convex lens to a photodiode detector. The light intensity is digitized with a DAQ device and stored on a personal computer. The spatiotemporal scanning is realized with a motorized stage controlled with the PC. The whole apparatus was placed in a warm room (37 °C) throughout the experiment

a direct spatial–temporal optical density measurement setup applied to Petri dish that provides quantitative and spatial–temporal cell density measurement (Fig. 2). Normally bacteria optical density measurement is carried out with run of the mill spectrophotometer and standard plastic cuvette with precise 1 cm optical path length. We utilize similar principle but with extensive calibrations to control for sources of systematic errors.

2 Materials

2.1 Reagent and General Facility

1. Engineered *E. coli* K12 MG1655 strain or other strains to be studied.

2. Semisolid agar plates: 10 ml of Luria–Bertani (LB) medium buffered by 0.1 M HEPES (pH 8.0) containing 0.25 % agar was poured into a 10-cm *Petri* dish and allowed to harden at room temperature for 90 min.

3. Paper template for marking the outline and center of 10 cm Petri dish for consistent seeding of *E. coli* at the center of each semisolid agar plate.

4. A room with constant temperature of 37 °C (*see* **Note 1**).

5. Blackout cloth to insulate ambient light.

6. MATLAB software for data analysis (Mathworks, Natick, MA, USA).

2.2 Bacteria Pattern Imaging

1. Camera and lens: Canon 450D digital single lens reflex camera and EF 50 mm f/1.8 II lens (*see* **Note 2**).

2. Tripod: Benro C-2691 carbon fiber tripod with B1 Ballhead.

3. Camera controller: Canon timer remote controller TC-80N3 for time-lapse image capturing.

4. Illumination and controller: Off-the-shelf LED light belt with 30 0.15 W white light-emitting diodes (LED) per meter, and 12 V regulated power supply of your trust brand.

5. Spirit level: Any brand from hardware stores.

6. Sheet materials for illumination box construction: 5 mm thick transparent polyacrylic sheet from local home improvement stores, professionally cut to four 14 cm-by-40 cm side panels and two 41 cm-by-41 cm top and bottom panels.

7. Light diffuser for LED light belt: Semiopaque plastic tubing with diameter of 5 mm.

2.3 Bacteria Cell Density Measurement in Agar Plates

1. One simple optical table with the size of 60 cm by 60 cm or bigger.

2. Motorized stage: Stand-alone LabVIEW programmable motorized one-dimensional stage with at least 75 mm travel.

3. Red laser: 650 nm/2 mW laser diode, DA650-2-3 (Huanic Co., Xi'an, China) (*see* **Note 2**).

4. Voltage-regulated power supply for red laser diode: Any 5 V unit.

5. Pinhole: A thin metal sheet with one 300 μm diameter pinhole drilled/laser burnt through.

6. Red laser intensity detector: Amplified Si photodetector, PDA36A (Thorlabs, Newton, NJ, USA) (*see* **Note 2**).

7. Lens: Plano-convex lens, $f = 75$ mm, LA1208 (Thorlabs, Newton, NJ, USA).

8. Various optomechanical components and mirrors necessary for laser, pinhole, and photodetector fasten and light path arrangement (Thorlabs, Newton, NJ, USA).

9. Adjustable fasten mechanism to lock Petri dish onto the motorized stage.

10. Data acquisition: USB-based DAQ device capable of digitizing analog signal at 12 bit/10 kHz, model USB-6009 (National Instruments, Austin, TX, USA).

11. Personal computer, with LabVIEW programmable software (National Instruments, Austin, TX, USA).

3 Methods

3.1 Construction and Usage of a Pattern Formation Imaging System

A circular white LED array illumination of the Petri dish from below (Fig. 1). The light will not be directly collected by the camera. Instead, *E. coli* scatters the light. The scattered light, which is positively correlated with cell density, is captured by the camera. The procedure for construction and using of this system is described as follows:

1. Level the bottom acrylic sheet panel (41 cm by 41 cm) (*see* **Note 3**) on the floor at a low-traffic region of the warm room.

2. Cut a 1.1 m LED light belt piece (*see* **Note 3**), cover it with a semitransparent plastic diffuser tubing of the same length, connect the two electrode with 2 m insulated metal wire, and connect the other ends of the wire to a 12-V regulated power supply with correct polarity.

3. Fix the LED light belt with diffuser tubing to the bottom panel in a circular loop with about 35 cm diameter, and align its center with the center of the bottom acrylic panel (*see* **Note 4**).

4. Glue the top panel and four side panels of the acrylic sheet together to form a box cover. Remove the inner layer of the non-transparent protective paper of the top panel. Cut and remove a centered circle with diameter of 8.5 cm from the outer/upper paper protective layer to allow the Petri dish to be illuminated. The rest of the paper protective layers are left untouched to block the unnecessary light leaking.

5. Put the box cover exactly on top of the bottom panel with circular LED light belt. Level the top panel. It effectively becomes a special light box. Keep the power supply outside the light box.

6. Put a testing agar plate with faint pattern on top of the circular light-exposed area of top panel switch on the 12-V power supply (*see* **Note 5**). From the top of the agar plate, there should be little LED light directly observable, and the *E. coli* density variations (patterns) should be easily observable as only the light scattered by *E. coli* reaches the eyes (*see* **Note 6**). If not, adjust the diameter of the circular LED light belt and the height of the light box.

7. Position the digital camera right on top of the Petri dish with distance H approximately 50 cm using a tripod and flexible Ballhead. The back of the camera needs to be leveled using a spirit level. Secure the camera using the corresponding fasten knobs on the tripod and Ballhead.

8. Carefully adjust the focus through back LCD panel using the "live view" mode and 10× zoom option. Secure the focus ring of the lens using a piece of post tape (*see* **Note 7**).

9. Set the exposure to ensure best imaging quality. Set the white balance to match the color temperature of the LED light belt. A typical setup is ISO200, f/5.6, and exposure 0.6 s. It needs to be determined using in-camera meters.

10. Connect the camera remote controller to the camera, and set it to take pictures every 10 min. Ensure enough battery charge and storage space on the memory card before experiment.

11. Carefully position a newly seeded semisolid agar Petri dish to the exposed center of top panel of the light box (*see* **Notes 8–10**). Lay a ruler by the plate to provide scale in the acquired images.

12. A large piece of blackout cloth is used to cover the entire system, except for the camera remote controller and power supply, to prevent ambient light from interfering with image acquisition. Record the time of seeding, and start the time-lapse acquisition loops using the remote controller.

13. After experiment, stop the camera remote controller, and take out the memory card. Transport the images to a computer, and use MATLAB program to exam the images. Find one of the three channels (RGB) representing the highest dynamic range without saturation, and use it to generate grayscale imaging for further analysis. Determine the converting factor between pixel and millimeter manually using the image of the ruler (sample images in Fig. 3).

3.2 Construction and Usage of a Spatial–Temporal Optical Density Measurement Setup

The imaging system described above can only qualitatively record the cell density of the pattern. We constructed another optical setup to quantify the cell density profiles. The spatial–temporal optical density measurement setup is designed to measure the attenuation of a 650 nm laser beam after passing through an agar plate with the right angle (Fig. 2). The optical density (OD, absorbance) is proportional to the product of cell density and optical path length (agar thickness). Standard spectrophotometer uses standard cuvette to ensure 1 cm optical path length, so that cell density can be determined from OD readout directly. The semisolid agar in Petri dish (*see* **Note 8**) usually forms an inverse tapered shape with the lowest thickness at the center. To avoid such systematic error on the optical path length, we measured the location-dependent transmittances for known standard cell densities.

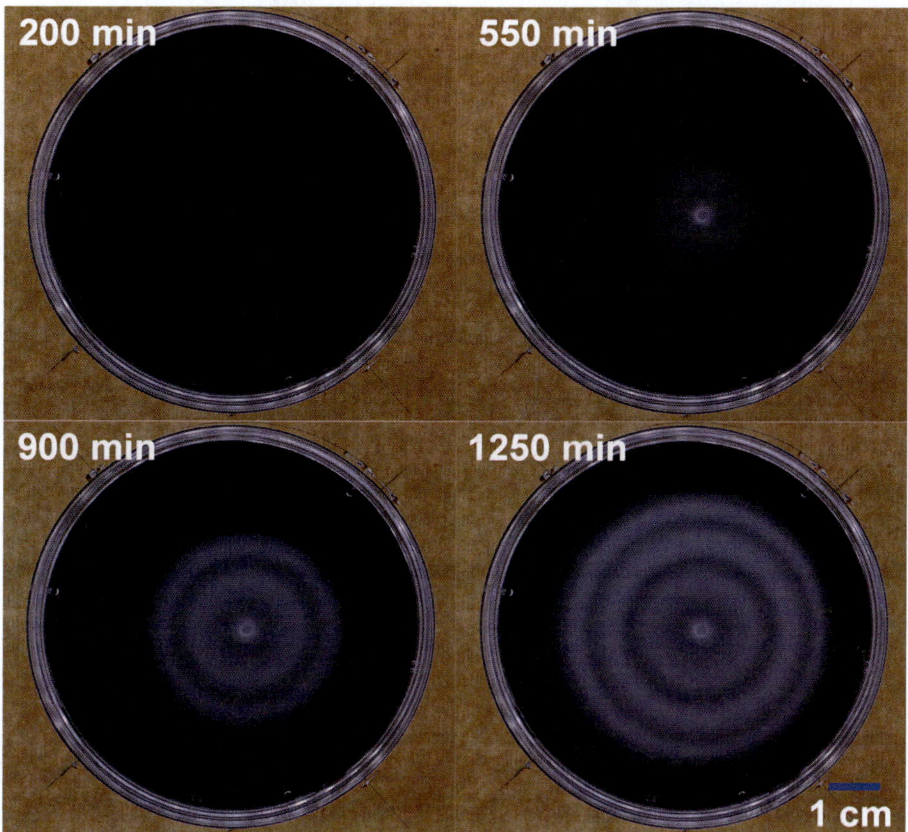

Fig. 3 Time-lapse pattern images of stripe-forming engineered *E. coli* using pattern formation imaging system. A series of images indicate the dynamics and geometry of the pattern formation

Hence we established a series of standard curves for any given radial positions. The step-by-step instruction is given below:

1. Position the optical table at the low-traffic area of the warm room.

2. Secure the motorized stage on the optical table, and level it using the spirit level (*see* **Note 11**).

3. Install a fasten mechanism for gently clamping the semisolid agar plate in place to the motorized stage, and clamp an empty and uncovered Petri dish for optical path adjustment (*see* **Notes 10** and **12**).

4. Fasten the 650 nm red laser diode to a kinematic mount, and position them right on top of the plate.

5. Wear proper laser safety goggle, and switch on the laser diode by connecting it to a 5-V regulated power supply.

6. Adjust the angle of the laser diode by adjusting the kinematic mount so that the reflection laser beam hits right back into the

front lens of laser diode. This ensures the right angle between the laser beam and agar plate.

7. Fine-tune the plate and laser beam so that the laser beam could scan right across the dead center of the plate with the motorized stage, and cover most of the radial space of the plate.

8. Position a thin sheet metal with 300 mm pinhole between the laser diode and the plate to generate a thin laser beam and improve the spatial resolution.

9. Position a collection convex lens below the plate to collect the transmission light.

10. Position an amplified photodiode detector so that all the collected transmission light hits the light-sensitive area of photodiode detector.

11. Connect the output of photodiode detector through a DAQ digitization device (*see* **Note 13**) to a computer (*see* **Note 5**), and visualize the light intensity using the accompanying Nation Instrument software for the DAQ. Set the digitization as 12 bit and 10 kHz.

12. Replace the empty plate with a blank semisolid agar plate (*see* **Note 14**).

13. Adjust the gain on the amplified photodiode detector so that, when scanning the empty semisolid agar, the signal collected by the detector and digitized by DAQ is as close to the maximum as possible without saturation. This ensures the maximum dynamic range of the system.

14. Connect the motorized stage through stage controller to the same computer. Write customized LabVIEW program to coordinate the stage movement, data acquisition through DAQ, and data storage. A diagram of the program is shown in Fig. 4.

15. Set the step size to 0.2 mm and total travel of 75 mm for the motorized stage to generate 376 radial positions.

16. At each position, the transmission light was sampled at 10 kHz and 12 bit using the DAQ device. The average of 1,000 data points was used to reduce the noise to the measurement at each position.

17. Set the repeat frequency for this scanning at once every 10 min.

18. Ensure the constant input laser intensity throughout the experiments with a regulated 5 V power supply, verified using the photodetector.

19. Cover the entire setup with blackout cloth using a metal frame to insulate the ambient light and prevent its interference with scanning.

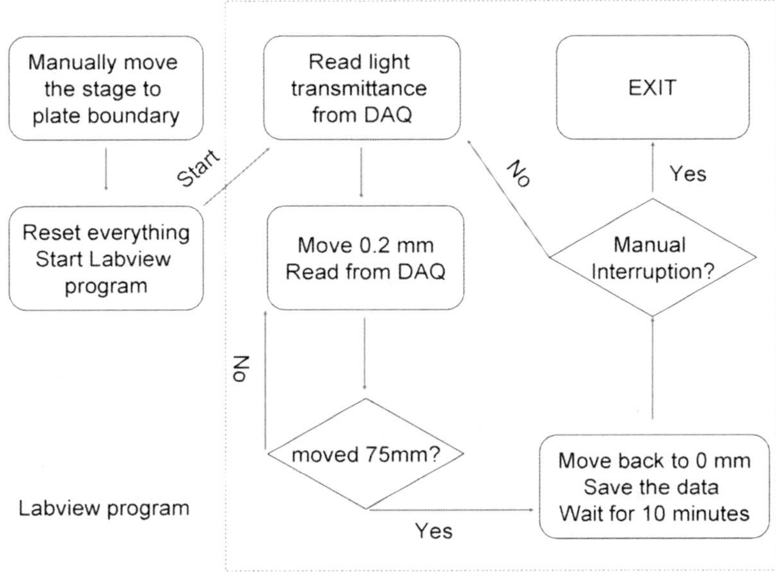

Fig. 4 A diagram of the automatic LabVIEW program performing periodic radial scanning of cell density profiles. The program is highlighted in the *dashed box*

20. The following **steps 21–29** are to obtain standard curves for each position for various known cell densities to calibrate this agar plate optical density scanner.

21. Grow a large amount of *E. coli* cells to mid-exponential phase (OD600 = 0.1–0.2).

22. Stop the cell growth by washing twice with nutrient-depleted LB and concentrate to 9.6×10^9 cells/ml.

23. Subsequently, serially dilute the cells with nutrient-depleted LB. For each cell density, vigorously mix 15 ml of cells with an equal volume of pre-warmed nutrient-depleted medium containing 0.5 % agar and pour into three Petri dishes with 10 ml each.

24. Allow all plates to harden at room temperature for 90 min. The final cell densities ranged from 0.03 to 4.8×10^9 cells/ml.

25. Move the plates into a warm room and place on the motorized stage one by one.

26. Scan the light transmittance radially for each plate, and load all the data using MATLAB program.

27. At each position along the radius, plot the ratio of the output intensities to the input intensities (measured using a blank agar plate), namely, the transmittances T, against the known cell densities for 21 different seeding densities (an example curve in Fig. 5a).

28. Generate totally 376 position-dependent density–transmittance standard curves (using a step size of 0.2 mm for a scanning

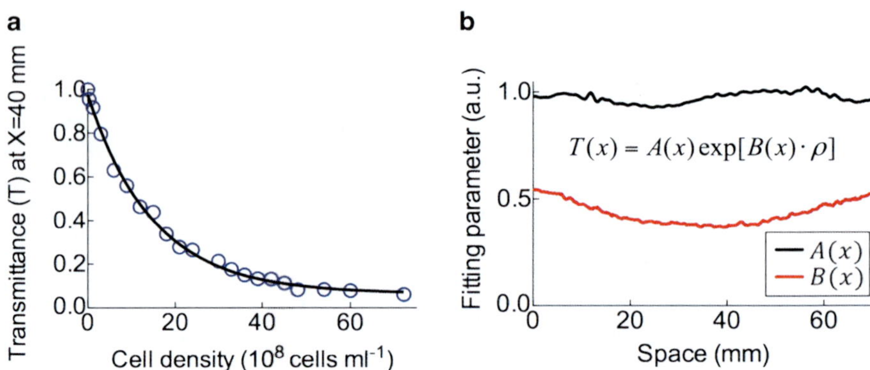

Fig. 5 The calibration curve of light transmittance versus real cell density in semisolid agar plates. (**a**) An example of the fitted curve of the transmittance as the function of cell density at a specific position. (**b**) The spatial distribution across the center of the dish, of the two fitting parameters for the standard curve, *A(x)* (*black line*) and *B(x)* (*red line*). This is derived from 376 standard curves (scanning range = 75 mm; step size = 0.2 mm)

range of 75 mm), and fit the results to an exponential function for each position *x*:

$$T(x) = A(x)\exp\left[B(x)\cdots\rho(x)\right],$$

where $A(x)$ and $B(x)$ are the position-dependent fitting parameters (Fig. 5b).

29. From these experimentally determined standard curves, we could invert them to compute the real cell density profile $\rho(x)$ from the measured transmittance profile $T(x)$.

30. For each experiment, place the semisolid agar plate seeded with engineered *E. coli* cell on the motorized stage, double check the scanning alignment, turn on the computer and power supplies for every component, cover with the blackout cloth, record the seeding time, and start the scanning.

31. After the experiments, convert the recorded light transmittance profiles $T(x)$ to the real cell density profile $\rho(x)$ for each time point.

32. For growth curve in semisolid agar, we uniformly mix the cell in semisolid agar plate, scan the transmittance periodically, and carry out the computation (see an example growth curve in Fig. 6). Growth parameters can be extracted and used in the mathematic models.

33. For measurement of cell density profile and dynamics, seed a small drop of engineered *E. coli* at the center of a semisolid agar plate, load it onto the setup, turn on the computer and power supply for laser diode and photodetector, load the LabVIEW program, cover the entire setup except for the computer, and start the periodic scanning (an example of such measurement in Fig. 7).

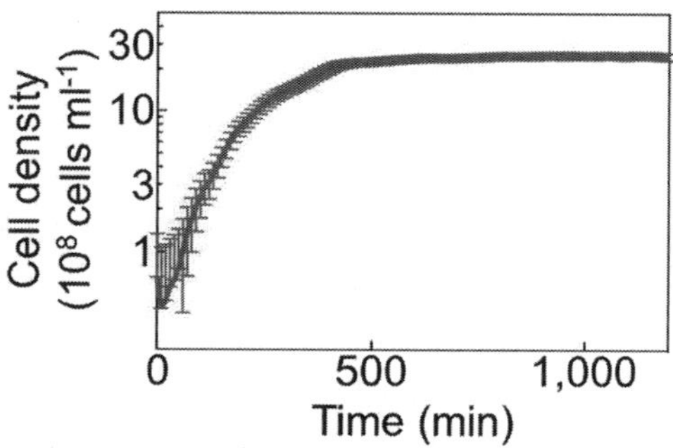

Fig. 6 A growth curve of *E. coli* in a Petri dish containing 10-ml 0.25 % LB agar. After seed culture and preculture growth, cells were diluted 200-fold into the pre-warmed LB media containing 0.25 % agar. Cell–agar mixtures were poured into Petri dishes, allowed to harden at room temperature for 90 min, and then moved back into a 37 °C incubator. At fixed intervals, cell density measurement is performed and computed

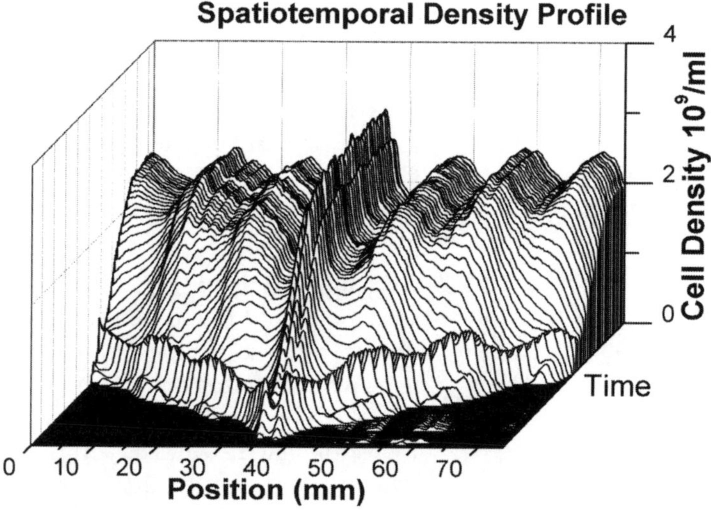

Fig. 7 Spatial–temporal cell density profiles provide quantitative characterization of the pattern-forming *E. coli*. This surface plot represents a typical stripe pattern

4 Notes

1. If the warm room is not available, it is possible to reduce the pattern formation imaging system to fit into a regular incubator. The optical density measurement system is too big to fit into any incubator. Therefore an environment chamber with adjustable temperature control will need to be built around the setup. Alternatively, one can implement this setup using a microscopy with environment chamber and motorized *x–y* stage.

2. With the optical setups running at 37 °C constantly, one biggest issue we faced is the reliability of the electrical and optical

components, as most of them are not designed for this temperature. We have narrowed done to the list of components after several rounds of trial and error, such as the laser, photodiode detector, and camera.

3. The dimension of the light box and circular LED light belt can be optimized manually prior to ordering the acrylic panel.

4. The light diffuser tube is not necessary if the LED density on the light belt is high enough. Colored LED should also be OK, but testing is needed.

5. The pattern-forming *E. coli* might be temperature sensitive. For the imaging system, the heat generated from the LED might heat up the light box significantly and alter the temperature of the agar plate. We found that by lowering the voltage of power supply to the LED light belt, this heating effect can be greatly attenuated. Heat-dissipating holes can be drilled to the side panels of the light box to reduce the heating effect. The computer could heat up the area around optical density measurement setup. It can be either moved further away or using a lower power one, such as a laptop computer instead.

6. This scattering imaging is very sensitive to imperfection in the surface of the exposed top acrylic circle as well as the Petri dish. Care needs to be taken to avoid scratching them or accumulate dusts. Alternatively it can be cut away and replaced with a thin glass circle for easy maintenance.

7. Any digital single-reflex camera will work due to its large CCD sensor size and higher sensitivities. Prime lens has better optical quality than zoom lens and also better light collection. Some lens, when pointed down, cannot lock the focus properly. It can be solved by using either tape to stop the drifting or the closest focus so that no further drifting is possible. To increase the dynamic range of the images raw image files instead of jpg files can be stored. But it takes much larger storage space.

8. Special care needs to be taken to make the agar plates for ensuring even cooling down of the agar. The table for pouring the plates needs to be leveled. For every stack of plates, the bottom one has different temperature as it is closest to the table; so it is neglected.

9. Consistent seeding of the *E. coli* at the dead center of the plate helps to obtain consistent good measurements. We use a printout paper sheet to align the seeding for every plate.

10. Semisolid agar plate is very gentle, so it needs to be handled with care, including appropriate leveling, and gentle movement with motorized stage.

11. It is also possible to perform two-dimensional scanning of the optical density of the agar plate if motorized x–y stage is available.

It will take much longer time if the same step size for one-dimensional scanning. The scanning time should be kept significantly shorter than the cell doubling time and characteristic time scale for pattern formation. If it is not possible, then increasing step size or reducing the scanning area is required.

12. When clamping the agar plate on the motorized stage, it is critical to use minimal force, as any force might deform the plate and interfere with density measurement.

13. Electronic devices, such as DAQ, should be pressed against the stainless steel optical table to alleviate heat building up within the devices.

14. For every experiment, a couple of blank agar plates are needed to obtain the spatial blank reading.

Acknowledgement

This work is supported by a Research Grants Council of Hong Kong General Research Fund (No. 767711) and a University of Hong Kong Faculty of Medicine Development Fund.

References

1. Wolpert L (2007) Principles of development, 3rd edn. Oxford University Press, Oxford

2. Chuong CM, Richardson MK (2009) Pattern formation today. Int J Dev Biol 53(5–6): 653–658, doi: 082594 cc [pii] 10.1387/ijdb. 082594cc

3. Davidson EH, Erwin DH (2006) Gene regulatory networks and the evolution of animal body plans. Science 311(5762):796–800, doi:311/5762/796 [pii] 10.1126/science. 1113832

4. Held LI (1992) Models for embryonic periodicity. Monographs in developmental biology, vol 24. Karger, Basel

5. Kondo S, Miura T (2010) Reaction-diffusion model as a framework for understanding biological pattern formation. Science 329(5999):1616–1620, doi:329/5999/1616 [pii] 10.1126/science.1179047

6. Turing AM (1990) The chemical basis of morphogenesis. 1953. Bull Math Biol 52(1–2): 153–197, discussion 119–152

7. Muller P, Rogers KW, Jordan BM, Lee JS, Robson D, Ramanathan S, Schier AF (2012) Differential diffusivity of Nodal and Lefty underlies a reaction-diffusion patterning system. Science 336(6082):721–724, doi:science.1221920 [pii] 10.1126/science.1221920

8. Elowitz M, Lim WA (2010) Build life to understand it. Nature 468(7326):889–890, doi:468889a [pii] 10.1038/468889a

9. Liu C, Fu X, Liu L, Ren X, Chau CK, Li S, Xiang L, Zeng H, Chen G, Tang LH, Lenz P, Cui X, Huang W, Hwa T, Huang JD (2011) Sequential establishment of stripe patterns in an expanding cell population. Science 334(6053): 238–241, doi:334/6053/238 [pii] 10.1126/science.1209042

Chapter 11

Transcriptome Analysis of a Microbial Coculture in which the Cell Populations Are Separated by a Membrane

Kazufumi Hosoda, Naoaki Ono, Shingo Suzuki, and Tetsuya Yomo

Abstract

The microbial coculture of multiple cell populations is used to study community evolution and for bioengineering applications. The cells in coculture undergo dynamic changes because of cell–cell and cell–environment interactions. Transcriptome analysis allows us to study the molecular basis of these changes in cell physiology. For transcriptome analysis, it is essential that the cell populations in the coculture are harvested separately. Here, we describe a method for transcriptome analysis of a microbial coculture in which two different cell populations are separated by a porous membrane.

Key words Transcriptome analysis, Microbial coculture, Synthetic ecosystem, Cell culture insert, Membrane coculture

---

1 Introduction

The microbial coculture of multiple cell populations is useful for understanding the ecology and evolution [1–4] as well as for bioengineering applications, such as fermentation and bioprocessing [5–9]. In these cocultures, cell–cell interactions can dynamically change the phenotype of the cells. The transcriptome analysis of the different cell populations in a coculture is one of the most efficient strategies for understanding the molecular basis of the phenotypic changes that result from cell–cell interactions [10].

For transcriptome analysis, it is essential that the cell populations in the coculture are harvested separately. Cell culture inserts with a porous membrane are useful for separating not only eukaryotic cell populations [11, 12] but also bacterial cell population [10]. Porous membranes can be permeable to solutes but not to cells, depending on the pore size. A coculture separated by a porous membrane (membrane coculture) makes it possible to reveal whether physical contact is necessary for cell–cell interactions and, if not, to harvest cell populations in the coculture separately for transcriptome analysis [10].

Lianhong Sun and Wenying Shou (eds.), *Engineering and Analyzing Multicellular Systems: Methods and Protocols,*
Methods in Molecular Biology, vol. 1151, DOI 10.1007/978-1-4939-0554-6_11, © Springer Science+Business Media New York 2014

Here, we describe a method for transcriptome analysis of bacterial cells in a coculture using a cell culture insert with a porous membrane. Specifically, we used a coculture consisting of two different populations of *Escherichia coli* cells as a synthetic cooperative system. The membrane separation of the coculture using a commercially available cell culture insert for cell culture plates made it possible to harvest the two cell populations in the coculture separately and to extract the total RNA from each cell population. The total RNA extracted was then used in a DNA microarray. When converting the raw fluorescence data into gene expression levels, we used the finite hybridization model [13] to produce better quality data for gene expression. Next, we describe a simple method for microarray data analysis that will distinguish coculture-specific changes in the cell state from changes in the cell state that would occur regardless of coculturing, such as changes associated with growth phase shift. When analyzing evolved and ancestral populations, this method can also be used to distinguish evolutionary changes from changes that would occur independent of evolution.

2   Materials

Prepare all of the solutions for the molecular work using DNase/RNase-free water.

2.1 Membrane Coculture

1. Culture plates: 6-well cell culture insert companion plates and cell culture insert with a porous membrane. Several companies offer comparable products. We for example used high-density (10^8/cm^2) 0.4 μm pored translucent PET membrane cell culture inserts from BD Falcon (Franklin Lakes, NJ, USA).

2. Incubator: Shaking plate incubator.

2.2 Total RNA Preparation

1. Ice-cold ethanol containing 10 % (w/v) phenol: Dissolve crystalline phenol in 99.5 % ethanol. Keep on ice before use.

2. RNeasy Mini Kit (Qiagen, P/N 74104).

3. RNase-free DNase Set (Qiagen, P/N 79254).

4. TE buffer: 10 mM Tris–HCl, 1 mM EDTA, pH 8.0.

5. Lysozyme in TE buffer: Dissolve lysozyme at 1 mg/mL in TE buffer.

6. β-Mercaptoethanol to Buffer RLT (Qiagen RNeasy Mini Kit) for a final concentration of 1 % (v/v) just before use.

7. Spectrophotometer.

2.3 DNA Microarray

1. dNTP solution (10 mM).

2. Random Primers (Life Technologies, P/N 48190-011).

3. GeneChip® Eukaryotic Poly-A RNA Control Kit (Affymetrix, P/N 900433).

4. SuperScript II™ Reverse Transcriptase (Life Technologies, P/N 18064-071).

5. SUPERase·In™ (Ambion, P/N 2696).

6. NaOH, 1 M solution.

7. HCl, 1 M solution.

8. MinElute PCR Purification Kit (Qiagen, P/N 28004).

9. Deoxyribonuclease I.

10. 10× DNase I buffer: 100 mM Tris–acetate, pH 7.5, 100 mM magnesium acetate, and 500 mM potassium acetate.

11. Agarose.

12. 5× TBE buffer: 445 mM Tris–borate and 10 mM EDTA (titrated to pH 8.0 with NaOH).

13. 1× TBE running buffer: Prepare from the 5× TBE buffer before use.

14. 50 bp DNA Step Ladder Marker.

15. 10× Loading buffer: 50 % glycerol, 0.9 % sodium dodecyl sulfate, and 0.05 % bromophenol blue.

16. SYBR Gold (Molecular Probes, P/NS-11494).

17. GeneChip® DNA Labeling Reagent (Affymetrix, P/N 900542).

18. Terminal deoxynucleotidyl transferase.

19. (Optional) ImmunoPure NeutrAvidin (Pierce Chemical, P/N 31000).

20. (Optional) PBS, pH 7.2: 1.54 mM Potassium phosphate monobasic, 155 mM sodium chloride, and 2.71 mM sodium phosphate dibasic.

21. (Optional) 4–20 % TBE Gel, 1.0 mm, 15 well (Life Technologies, P/N EC62252).

22. Affymetrix GeneChip® *E. coli* Antisense Genome Array (Affymetrix, P/N 510052).

23. GeneChip® Hybridization, Wash and Stain Kit (Affymetrix, P/N 900720).

24. Control Oligo B2, 3 nM (Affymetrix, P/N 900301).

25. Hybridization Oven 640 (Affymetrix, P/N 800138).

26. Fluidics Station 450 (Affymetrix, P/N 00-0079).

27. GeneChip® Scanner 3000 (Affymetrix, P/N 00-00212).

2.4 Data Conversion from Raw Fluorescence to Gene Expression Level

1. FHarray package [13] that is specific to the open-source R software [14] (*see* Subheading 3.4 for its download and installation).

2.5 Analysis of the Transcriptome Data

1. Basic computational software, such as MATLAB (MathWorks, MA, USA), R, or Excel (Microsoft, WA, USA). Here, we show example source cords in MATLAB as one of the most commonly used platforms. Statistics Toolbox and Bioinformatics Toolbox are required for MATLAB to use the code described here.

3 Methods

Conduct all of the inoculations and incubations under sterile conditions and all of the molecular work under DNase/RNase-free conditions.

3.1 Membrane Coculture

Prepare two isolated populations (*see* **Note 1**).

1. Set a cell culture insert in a well of the culture plate.

2. Inoculate the two different cell populations separately: 1,750 µL of one population above the insert and 2,250 µL of the other population below the insert (Fig. 1a).

3. Incubate the coculture with (450 rpm, *see* **Note 2**) or without shaking.

4. Sample the cell solutions to measure the cell concentration (Fig. 1b; *see* **Note 3**). At this time, determine whether the cells can pass through the membrane (*see* **Note 4**).

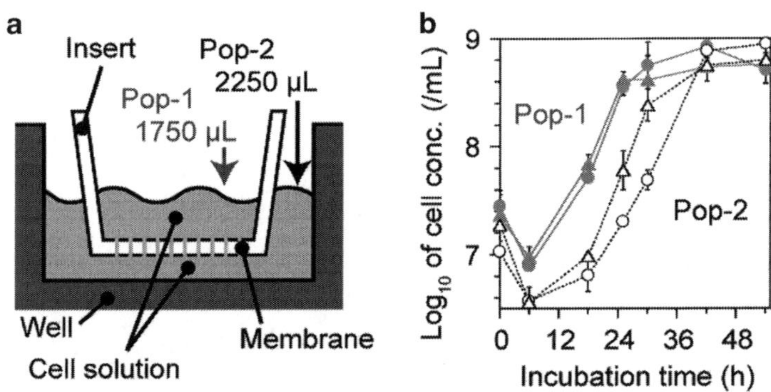

Fig. 1 The membrane coculture. (**a**) A schematic diagram of the inoculation of two different cell populations (Pop-1 and Pop-2) is shown. (**b**) Example results of growth curves for a membrane coculture. *Filled-gray* and *open-black* symbols represent two different cell populations, Pop-1 and Pop-2, respectively. *Circles* and *triangles* indicate results from a coculture with and without membrane separation, respectively. We used flow cytometry to measure cell concentrations, which made it possible to distinguish the two populations in the coculture without a membrane because we used two *E. coli* populations that were fluorescently labeled in two different colors (DsRed.T3 and GFPuv5). From these results, one can evaluate the effect of physical contact on the cell population and can harvest cell populations separately for transcriptome analysis if physical contact is negligible for the purpose of the study

3.2 Total RNA Preparation

1. Harvest the cell suspensions from above and below the insert in the membrane coculture separately.

2. Immediately add an equal volume of ice-cold ethanol containing 10 % (w/v) phenol to each cell solution, and mix it gently but rigorously.

3. Centrifuge the solutions for 5 min at maximum speed (>10,000×g) and 4 °C.

4. Carefully remove and discard the supernatant, and store the cell pellets at −80 °C until needed.

5. Add 100 µL of lysozyme (1 mg/mL) in TE buffer to each cell pellet, and pipet up and down several times.

6. Incubate the lysozyme reactions at room temperature for 5 min.

7. Add 350 µL of buffer RLT (RNeasy Mini Kit) containing 1 % (v/v) β-mercaptoethanol, vortex the tubes, and pulse-spin the tubes in a centrifuge to collect the samples.

8. Add 250 µL of 99.5 % ethanol, and pipet up and down several times. Do not pulse-spin the tubes.

9. Load onto RNeasy Mini columns, and purify the total RNA using an RNeasy Mini Kit with on-column DNA digestion in accordance with the manufacturer's instructions.

10. Measure the total RNA concentration using a spectrophotometer.

11. (Optional) Assess the total RNA quality by gel electrophoresis (*see* **Note 5**).

3.3 DNA Microarray

The total RNA prepared here can be used on any DNA microarray platforms. Follow the manual of the platform you use. We summarize here steps for the GeneChip® system from Affymetrix with *E. coli* antisense genome array as an example (catalog number 510052, Affymetrix).

1. cDNA synthesis: Aliquot 10 µg total RNA, and follow the protocols for the 49 format array in Chapter 4, "Prokaryotic Target Preparation," of the GeneChip® Expression Analysis Technical Manual (Affymetrix). The single-strand cDNA yield should be approximately 4–5 µg.

2. cDNA fragmentation: Aliquot 4 µg of single-strand cDNA, and follow the protocol in Chapter 4, "Prokaryotic Target Preparation," of the GeneChip® Expression Analysis Technical Manual (Affymetrix).

3. Examination of the cDNA fragmentation: Prepare a 3 % (w/v) agarose gel in 1× TBE buffer in accordance with the manufacturer's instructions. Load 200 ng of cDNA on the gel, and electrophorese at 100 V for 35 min. Make sure to add 10×

DNaseI buffer to the DNA ladder marker. Stain the gel with SYBR Gold. The majority of the fragmented single-strand cDNA should be in the 50–200 base range.

4. Terminal labeling of the fragmented cDNA: Aliquot 3.5 µg of fragmented cDNA, and follow the protocol in Chapter 4, "Prokaryotic Target Preparation," of the GeneChip® Expression Analysis Technical Manual (Affymetrix).

5. (Optional) Assess the efficiency of the labeling procedure using a gel-shift assay in accordance with the protocol in Chapter 4, "Prokaryotic Target Preparation," of the GeneChip® Expression Analysis Technical Manual.

6. Microarray hybridization: Aliquot 3 µg of terminal-labeled cDNA, and follow the protocol in Chapter 5, "Prokaryotic Target Hybridization," of the GeneChip® Expression Analysis Technical Manual (Affymetrix).

7. Microarray washing, staining, and scanning: Follow the protocol in Chapter 6, "Prokaryotic Arrays: Washing, Staining and Scanning," of the GeneChip® Expression Analysis Technical Manual (Affymetrix).

8. Export your data to a file (CEL format).

3.4 Data Conversion from Raw Fluorescence to Gene Expression Level

The expression level of each gene is determined from the raw fluorescence intensity data using the finite hybridization model (FH model [13]) package specified for R software. In general, the expression level of each gene is calculated based on the statistical evaluation of a set of signal values for the DNA probes designed for that gene on the microarray. However, because of the nonlinear relationship between the gene expression level, i.e., the concentration of cDNA, and the fluorescence intensity, the dynamic range of the quantitative estimates was limited to a certain range. To increase this dynamic range for evaluating the expression levels in the FH model, a detailed physicochemical model of the hybridization between the DNA probes on the chip and cDNA fragments in the sample was proposed (Fig. 2). To improve accuracy, the hybridization affinity of every probe was predicted based on its sequence and according to the parameters that were optimized using the data from RNA titration samples as a control (*see* **Note 6**). To reduce background noise, this model also predicts the amount of cross hybridization. An expression analysis using this model can quantify cDNA concentration levels spanning over five orders of magnitude (*see* **Note 7**). Here we describe an R command sequence for the conversion from CEL files to a gene expression data matrix, using *E. coli* antisense genome array as an example (*see* **Note 6**). Below, all R commands after ">" need to be typed in, where ">" is prompted automatically by R ("+" is also prompted automatically when the command is separated by a line break and incomplete).

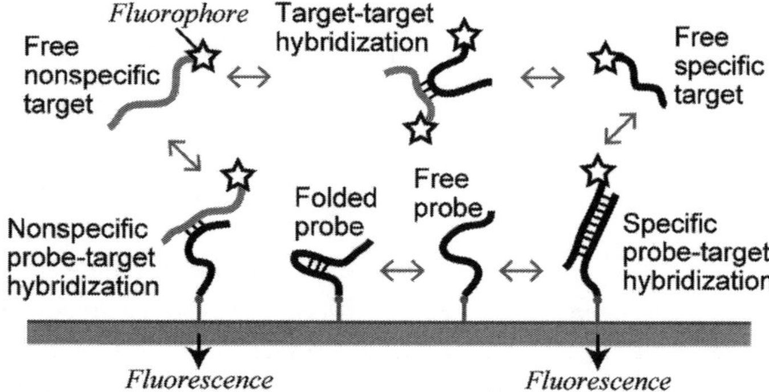

Fig. 2 Scheme of the finite hybridization model. This model assumes that the DNA probes immobilized on chip surface and cDNA fragments prepared from the sample exist in the depicted states. The free energy differences between these states are estimated from their sequences, and the amount of target molecules on the microarray surface in the equilibrium is computed to predict fluorescence intensity

1. Install R software to your computer after downloading the R package from http://www.r-project.org/.

2. Preparation before the installation of the FHarray package: From the "Packages" tab of R, set "repositories" to "Bioconductor (BioC software)" and install "Biobase" and "affy" packages.

3. Download the FHarray package (FHarray_xxx.tar.gz, xxx is the version number) from http://www-shimizu.ist.osaka-u. ac.jp/shimizu_lab/FHarray/.

4. Install the FHarray package by the command

```
>    install.packages("fp",    type="source",
repos=NULL)
```

   where *fp* is a file path (e.g., C:/Users/FHarray_xxx.tar.gz) that you want to save the package to.

5. Load the FHarray package using the command

```
> library(FHarray)
```

   This will load FHarray as well as the Biobase and affy packages and their dependencies.

6. Data import: First, set the working directly by the command

```
> setwd("dp")
```

   where *dp* is a directory path, where the data files (CEL format) exist. Then, read your data file by the command

```
> cell<- read.affybatch("fn")
```

   where *fn* is a file name of your data. You may have a warning message "Incompatible phenoData object. Created a new one." but this is not a problem. Since you would have multiple files for comparison, use comma to separate files, e.g., ("*fn₁*", "*fn₂*","*fn₃*").

7. Prepare probe information: Obtain probe set list from CEL data by command:

```
> index.EcASv2<- indexProbes(cel1)
> probesetnames.EcASv2<- names(index.EcASv2)
```

Choose probe sets to analyze. In this example, probe sets whose names include gene number (bxxxx, which is specific for *E. coli*) were selected by the command

```
>    index.orf<-    grep("b[0-9]{4}",probeset-
names.EcASv2)
```

where the function grep searches for matches to argument pattern (here, bxxxx, x is a number) within each element of a character vector (here, probesetnames.EcASv2). Choose probe sets such as Affymetrix's control probes whose target genes are not contained in the sample to evaluate background effect by the command

```
> index.notarget<- grep("AFFX",
+   probesetnames.EcASv2)[c(3:4,8:10,20:22)]
```

Load parameter data from the library by the command

```
> data(parameterWithBG)
```

To import probe information from a probe table file, which is provided for each type of probe array chips by Affymetrix at http:www.affymetrix.com/support/technical/byproduc.affx?cat=arrays, type

```
> probeTable.EcASv2<- read.delim("pt",
+   stringsAsFactors=FALSE,header=TRUE,sep=
"\t")
```

where *pt* is the file path of the probe table file.

8. Compute gene expression level by the command

```
> expression1<- evaluateProbeSet(index.orf,
+   cel1,probeTable.EcASv2,parameterWithBG,
+   NA,index.notarget)
```

You can see that the data matrix "expression1" is a log-scaled (base 10) gene expression where rows and columns correspond to genes and samples, respectively. For example, the entry in the first row and second column corresponds to the intensity of Gene 1 in Sample 2. This data matrix is treated as X in the following section (*see* Subheading 3.5).

9. Export the gene expression to a file by the command

```
> write.csv(expression1, file="export.csv")
```

3.5 Analysis of the Transcriptome data

Transcriptome analyses require data mining. Here, we describe simple analyses to extract coculture-specific changes (*see* **Note 8**) that require basic software, such as MathWorks MATLAB, GNU R, or Microsoft Excel.

As an example, we show commands in MATLAB, assuming that we have the log-scaled (usually base 10) gene expression data matrix X where rows and columns correspond to genes and samples, respectively (such as obtained from Subheading 3.4). Below, all MATLAB commands after ">>" need to be typed in, where ">>" is prompted automatically by MATLAB.

1. Normalization 1 "Scaling normalization": First, for each sample, find the median value of log(expression). Subtract this median from each log(expression) within the sample to obtain matrix Y. In MATLAB, use the command

```
>>Y=X - ones(length(X),1) * median(X)
```

   We can then visualize the frequency distribution of expression values of all samples, with the median of each sample being zero as expected (Fig. 3a). In MATLAB, use the command

```
>>bin=-5:0.02:5
>>plot(bin', hist(Y,bin))
```

2. (Optional) Normalization 2 "Quantile normalization [15]": Sort the expression levels in decreasing order for each column independently, and calculate the average value in all samples for every rank in the order. Then, replace the original value in each sample with the average value for that same rank. This allows all samples to have the same value for the same rank (but not for the same gene). In MATLAB, use the command

```
>>Z=quantilenorm(Y)
```

   Note that you need Bioinformatics Toolbox in MATLAB for the command. This step converts the distribution curves for the expression levels of all of the samples to be the same. This step should be taken if you need to equally treat the scale of changes in gene expression in samples that have different distributions (*see* **Note 9**). Hereinafter, Z instead of Y is analyzed.

3. Principal component analysis (PCA): Conduct PCA, and plot the main characteristics of samples (Fig. 3b shows the difference in the populations and the growth conditions; *see* **Note 10**). In MATLAB, use the command

```
>>mapcaplot(Y')
```

   Note that you need Bioinformatics Toolbox for the command.

4. Gene scatter plot 1: A simple way to compare two samples is a scatterplot of the log-scaled expression levels to find significant differences (Fig. 3c; *see* **Note 11**). For example in MATLAB, to compare sample in the first column with sample in the second column with each dot plotted as black (k) dot (.), use the command

```
>>plot(Y(:,1),Y(:,2),'k.')
```

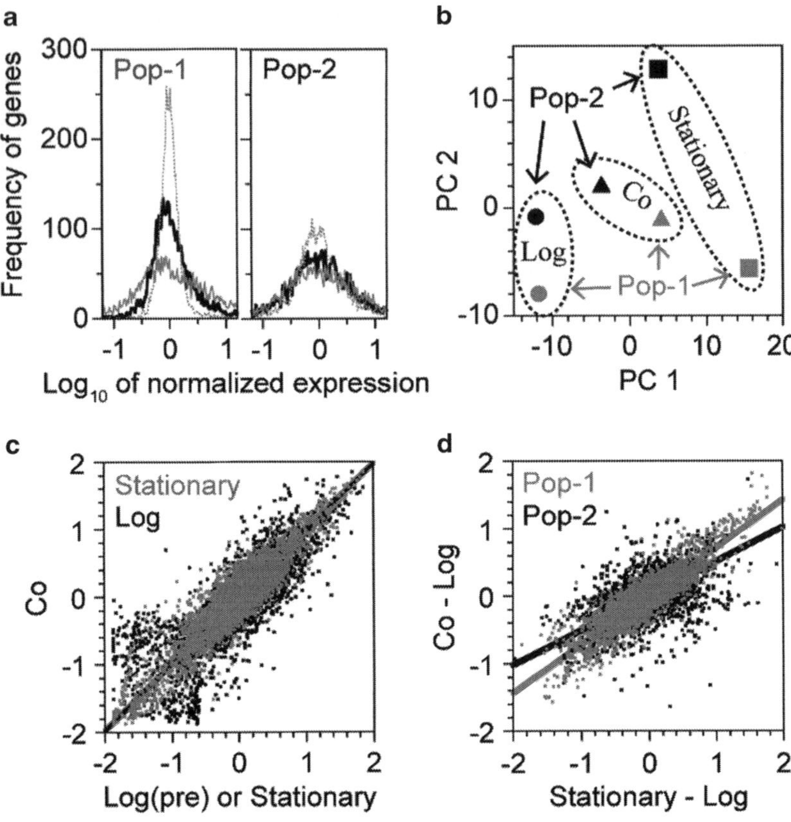

Fig. 3 Example results of data analyses. (**a**) An example of gene frequency distributions of the normalized expression level. The two panels depict the two different cell populations, Pop-1 and Pop-2. The *black-bold, gray-solid,* and *gray-dotted lines* indicate data from the coculture, log-phase monoculture, and stationary-phase monoculture, respectively. The shape of the distribution curve depends on the growth condition and cell population. (**b**) Example results of a PCA. The positions of points indicate the abstract state of the cell sample; closer points would be in more similar cell states. *Gray* and *black* indicate Pop-1 and Pop-2, respectively. The *triangle, circle,* and *square* indicate the data for the coculture, log-phase monoculture, and stationary-phase monoculture, respectively. (**c**) Example results of a gene scatterplot 1. One dot represents the expression of one gene in two different samples. This plot shows the expression comparison of cells in coculture versus cells at log phase in monoculture (i.e., preculture, black dots) or versus cells at stationary phase in monoculture (*gray dots*) of population 1. We can see strong correlations (with the slope of nearly 1) in both comparisons. (**d**) Example results of a gene scatterplot 2. The *gray* and *black* dots indicate Pop-1 and Pop-2, respectively. The *horizontal axis* is the change between stationary-phase monoculture and log-phase monoculture. The *vertical axis* is the change between coculture and log-phase monoculture. This plot more clearly shows the coculture-dependent changes because gene-specific expression scale is normalized by the log-phase monoculture (preculture). We can see that the correlation slope was less in Pop-2 than in Pop-1, which suggests that Pop-1 in coculture was closer to stationary phase than Pop-2. We can also see that the deviation from the regression line was greater in Pop-2 than in Pop-1, which suggests that Pop-2 was more influenced by coculture than Pop-1

5. Gene scatterplot 2: We describe a simple and effective method for distinguishing between coculture-specific changes and changes associated with growth phase shift (such as entering stationary phase) regardless of coculturing. Determine the

change in cell state from before and after coculturing using the difference in the expression data of the "coculture − preculture" in log scale. Next, determine the change in cell state between different growth conditions (for example, if the cells were at the log phase in the preculture, then the "stationary phase in monoculture − preculture" can be used). Plot these changes (Fig. 3d). In MATLAB, if the columns 1, 2, and 3 correspond to samples "at the stationary phase in monoculture," "in coculture," and "at the log phase in preculture," respectively, use the command

```
>>plot( Y(:,1)-Y(:,3) , Y(:,2)-Y(:,3) )
```

A slope close to 1 suggests that the state is close to the state of comparison, in this case, stationary. Deviation from the regression line suggests coculture-specific changes (*see* **Note 12**). The difference between Fig. 3c (gray) and 3d (gray) is just the normalization by the subtraction of the log-phase preculture, and Fig. 3d shows coculture-specific changes more efficiently.

6. Gene screening: Statistically screen the genes whose changes were specific to coculture. For example, calculate the distribution of the residual value from the regression line. If the distribution can be approximated by a normal distribution (in log scale), the genes that changed significantly can be screened using a statistical test. For example, you can calculate the z value of a Z-test for each gene by calculating $z=(x-\mu)/\sigma$, where x is the residual value of a gene and μ and σ are mean and standard deviation of the all sample (approximately 4,000 genes in the case of *E. coli*), respectively. You can estimate the p-value of the statistical test (p) from the z value for each gene. When you compare p and the significance level α, you should perform a multiple-comparison correction on α. For example, you can obtain a corrected significance level as $\alpha_{\mathrm{cor}}=\alpha/n_{\mathrm{g}}$, where n_{g} is the number of total comparisons which in our case is the number of total genes (this correction is one of the simplest methods called Bonferroni correction [16]).

7. Functional analysis: Screen genes with similar functions to summarize the changes in transcriptome in terms of cellular functions in order to find functional categories in which many up/downregulated genes are included. For example, for each given functional categories, such as GO terms [17], use a binomial or a hypergeometric test to evaluate whether the ratio of up/downregulated genes in each category is significantly larger than the ratio of up/downregulated genes in all genes (total ratio). This step assumes that genes with similar functions share some regulatory mechanisms. On the other hand, clustering methods, such as hierarchical clustering and a self-organizing map, could also be used to find clusters of genes that showed characteristic expression changes. For example, for hierarchical clustering, use the command

```
>>clustergram(Y)
```

Note that you need Bioinformatics Toolbox in MATLAB for the command. You may screen clusters where the ratio of included genes of some specific functions is significantly larger than the total ratio.

4 Notes

1. If you have experimentally evolved populations, then you could use this technique to investigate the coculture of an ancestral and an evolved population as well as the coculture of a pair of ancestral populations and the coculture of a pair of evolved populations. This membrane coculture system should only be used to investigate cells in coculture that hardly depend on any physical contact between the cell populations. Taking this caveat into consideration, you could use this membrane coculture to investigate whether physical contact between the cell populations is negligible.

2. Shake the culture using the highest speed possible but not at a speed that would allow the culture solution to splash or spread out from the well and insert. Verify that cells are indeed mixed by ensuring that samples from different locations have similar cell densities.

3. You can use any measurement method, such as optical density. We recommend, however, that you use a method that can immediately distinguish between the populations, such as a difference in fluorescence color or cell size, because you must check whether the cells can pass through the membrane (*see* **Note 4**).

4. Cells can pass through the membrane depending on the cell size and the membrane pore size. In our case using *E. coli*, the cells did occasionally pass through the insert (approximately one out of four cases in a 1-day culture). Therefore, we cultivated six independent membrane cocultures, measured the cell concentration, and selected the "unmixed" sample for further analysis. We would like to emphasize that we used a high-density porous insert for the coculture (*see* Subheading 2.1). When we used a low-density porous insert, the cells passed more frequently (we do not know the cause). You may try another insert. Furthermore, we rotary shook the membrane coculture and never tried a static culture.

5. The intensity ratio for the 23S and 16S rRNA should be approximately 1.8.

6. Since the model was optimized using *E. coli* data, it may improve the accuracy of quantification to re-optimize the nearest-neighbor parameters used in the FH model when you use

other types of arrays, as follows: (1) Choose sets of control probes whose targets are not included in the sample transcriptome (e.g., Affymetrix's control probes). (2) Prepare the control RNAs by in vitro transcription. (3) Add the control RNAs at various known concentrations to an RNA sample prepared from a representative cell extract, and measure their fluorescence intensity. Since the concentrations of the control RNAs are known, the predicted intensity $\mathbf{I}_{pre}$ can be computed as a function of model parameters. Consider the residual error RE between the predicted and observed intensity values as

$$\mathrm{RE}(\mathbf{p}) = \frac{1}{N}\sum\left(\log\mathbf{I}_{obs} - \log\mathbf{I}_{pre}(\mathbf{p})\right)^2,$$

where $\mathbf{I}_{obs}$ represents the observed probe intensity of the control RNA, $\mathbf{p}$ represents a vector of model parameters, and N is the number of corresponding probes of the control RNA fragments. (4) Optimize the model parameters to minimize the residual error. General optimization methods (many of them are implemented in the R software) can be applied to obtain the values of optimized model parameters.

7. You may use another model, such as the one that comes with the instrument. However, we recommend using the FH model for the data conversion because the FH model dramatically increases (around two orders of magnitude) the cDNA concentration range that can be estimated (see [13] for more detail).

8. If you have evolved populations, you may be able to investigate the evolutionary changes using this same logic.

9. Quantile normalization (frequently used to determine the distribution of raw probe intensities [15]) changes the expression values of each gene depending on the whole set of expression data, which is unsuitable when you need to compare the values themselves.

10. If you have an evolved population, this method should reveal an evolutionary difference.

11. You may be able to plot an ancestral population versus an evolved population using this method.

12. You may be able to determine the evolutionary changes using this same logic.

Acknowledgements

This work was supported in part by JSPS KAKENHI grant number 25650147 and the "Global COE Program" of the Ministry of Education, Culture, Sports, Science and Technology, Japan.

References

1. Shou W, Ram S, Vilar JM (2007) Synthetic cooperation in engineered yeast populations. Proc Natl Acad Sci U S A 104:1877–1882

2. Wintermute EH, Silver PA (2010) Dynamics in the mixed microbial concourse. Genes Dev 24:2603–2614

3. Momeni B, Chen CC, Hillesland KL, Waite A, Shou W (2011) Using artificial systems to explore the ecology and evolution of symbioses. Cell Mol Life Sci 68:1353–1368

4. Tanouchi Y, Smith RP, You L (2012) Engineering microbial systems to explore ecological and evolutionary dynamics. Curr Opin Biotechnol 23(5):797–797

5. Brenner K, You L, Arnold FH (2008) Engineering microbial consortia: a new frontier in synthetic biology. Trends Biotechnol 26:483–489

6. Chen Y (2011) Development and application of co-culture for ethanol production by co-fermentation of glucose and xylose: a systematic review. J Ind Microbiol Biotechnol 38:581–597

7. Hosoda K, Yomo T (2011) Designing symbiosis. Bioeng Bugs 2:338–341

8. Shong J, Jimenez Diaz MR, Collins CH (2012) Towards synthetic microbial consortia for bioprocessing. Curr Opin Biotechnol 23:798–802

9. Mee MT, Wang HH (2012) Engineering ecosystems and synthetic ecologies. Mol Biosyst 8:2470–2483

10. Hosoda K, Suzuki S, Yamauchi Y, Shiroguchi Y, Kashiwagi A, Ono N et al (2011) Cooperative adaptation to establishment of a synthetic bacterial mutualism. PLoS One 6:e17105

11. Halleux C, Schneider YJ (1991) Iron absorption by intestinal epithelial cells: 1. CaCo2 cells cultivated in serum-free medium, on polyethylene terephthalate microporous membranes, as an in vitro model. In Vitro Cell Dev Biol 27A:293–302

12. Dehouck MP, Meresse S, Delorme P, Fruchart JC, Cecchelli R (1990) An easier, reproducible, and mass-production method to study the blood-brain barrier in vitro. J Neurochem 54:1798–1801

13. Ono N, Suzuki S, Furusawa C, Agata T, Kashiwagi A, Shimizu H et al (2008) An improved physico-chemical model of hybridization on high-density oligonucleotide microarrays. Bioinformatics 24:1278–1285

14. R Development Core Team (2006) R: a language and environment for statistical computing. R Foundation for Statistical Computing, Vienna, Austria

15. Bolstad BM, Irizarry RA, Astrand M, Speed TP (2003) A comparison of normalization methods for high density oligonucleotide array data based on variance and bias. Bioinformatics 19:185–193

16. Salkind NJ, Rasmussen K (2007) Encyclopedia of measurement and statistics. SAGE Publications, Thousand Oaks, CA

17. Keseler IM, Mackie A, Peralta-Gil M, Santos-Zavaleta A, Gama-Castro S, Bonavides-Martinez C et al (2013) EcoCyc: fusing model organism databases with systems biology. Nucleic Acids Res 41:D605–D612

Identification of Mutations in Laboratory-Evolved Microbes from Next-Generation Sequencing Data Using *breseq*

Daniel E. Deatherage and Jeffrey E. Barrick

Abstract

Next-generation DNA sequencing (NGS) can be used to reconstruct eco-evolutionary population dynamics and to identify the genetic basis of adaptation in laboratory evolution experiments. Here, we describe how to run the open-source *breseq* computational pipeline to identify and annotate genetic differences found in whole-genome and whole-population NGS data from haploid microbes where a high-quality reference genome is available. These methods can also be used to analyze mutants isolated in genetic screens and to detect unintended mutations that may occur during strain construction and genome editing.

Key words Evolutionary genomics, Genome re-sequencing, Variant caller, Single-nucleotide variant, Structural variant, Insertion sequence, Mobile genetic element, Gene conversion

1 Introduction

Recent developments in next-generation DNA sequencing (NGS) technologies have increasingly made them affordable and accessible to any researcher [1, 2]. The NGS platforms that are widely available today—including Illumina, Roche 454, Ion Torrent, Pacific Biosciences, and ABI SOLiD Systems—produce many DNA sequencing reads (thousands to millions to billions) of various lengths (~50 to >3,000 bases). This data is fundamentally different from classic Sanger sequencing, which generates just one read of 400–900 bases per reaction, in that NGS instruments generate many orders of magnitude greater base coverage, but each individual read sequence is often shorter and of lower quality. NGS technologies can be used for RNA-seq transcriptomics, the *de novo* assembly of non-model organism genomes, and characterizing epigenetic DNA modifications, among many other types of studies. One of the most common uses of NGS is to "re-sequence" samples from a laboratory organism or population that is very closely related (typically >99.9 % nucleotide identity) to a complete, high-quality reference genome to identify the salient genetic differences between them.

Lianhong Sun and Wenying Shou (eds.), *Engineering and Analyzing Multicellular Systems: Methods and Protocols,*
Methods in Molecular Biology, vol. 1151, DOI 10.1007/978-1-4939-0554-6_12, © Springer Science+Business Media New York 2014

There are three main steps in analyzing NGS genome re-sequencing data: (1) mapping each sequencing read to the reference genome, (2) identifying genetic variation present in the sample by searching for discrepancies between aligned reads and the reference genome, and (3) annotating how genes are affected by these sequence differences. Many software tools exist for read mapping, with various trade-offs in speed and sensitivity and algorithmic subtleties that can affect the downstream analysis steps [3]. Similarly, variant callers differ a great deal, both in how sophisticated their statistical models are for maximizing sensitivity while minimizing false-positive predictions and in what types of genetic variation they are designed to find [4, 5]. Three main categories of genetic variation exist: changes of a single base (single-nucleotide variants, SNVs), insertions and deletions of a few nucleotides (indels), and more complicated chromosomal rearrangements and larger insertions and deletions (structural variants, SVs). The latter types can be considerably more challenging to identify from NGS data. Many research groups and sequencing centers have created custom computational pipelines tailored to their needs by combining any number of read mapping, variant calling, and annotation programs.

Here, we describe how to use *breseq*, an open-source pipeline that automates all of the NGS genome re-sequencing analysis steps from mapping to annotation. In contrast to workflows developed for analyzing mainly human genomes [4, 5], *breseq* has been optimized for haploid microbial sized genomes (<20 Mb). Because *breseq* is intended for use on laboratory evolution experiments, molecular genetic experiments, and synthetic biology experiments with microbes—where detecting a single key genetic change in a sample can be very important [6–9]—it emphasizes sensitivity over speed and reports evidence for a wider variety of genetic variants than most other tools that are currently available. The *breseq* pipeline produces output in an annotated HTML format that is accessible to non-experts; in a Genome Diff flat file format for comparing mutations predicted in different samples and for applying mutations to a reference genome; and also in community formats that can be used for visualizing mapped reads and other downstream analyses.

2 Materials

2.1 Computer System and Software

1. Access to a computer system with a command-line prompt in a Unix-like environment, such as Linux or Mac OS X. On Windows machines, it is possible to compile and run *breseq* under Cygwin. For analysis of a few samples, a personal computer is likely sufficient. If you have many samples to

process, you may want to install and use *breseq* on a computer cluster or in a cloud computing environment.

2. *breseq* pipeline (http://barricklab.org/breseq): Download, compile, and install according to the included documentation. Version 0.24 was used to generate the figures and examples for this tutorial.

3. For editing Genome Diff files, a plain text editor that can be set not to wrap lines of text such as `vi` or `emacs` on a Unix-like system, TextWrangler, a Mac OS X system, or Notepad++ on a Windows system.

4. For viewing read alignments, Tablet (http://bioinf.scri.ac.uk/tablet/) [10] or the Integrative Genomics Viewer (http://www.broadinstitute.org/igv/) [11].

2.2 Data and Tutorial Files

1. NGS read files for clonal or whole-population genomic DNA samples in FASTQ format. These files should be provided by the facility that sequenced your samples. *breseq* does not require input FASTQ files to use a specific base quality encoding scheme, and it is compatible with most current technologies, except SOLiD color space data.

2. Reference genome sequence files in GenBank, GFF3, or FASTA format. GenBank or GFF3 files with feature annotations are preferred, as they enable *breseq* to report the effects of predicted mutations on genes. Suitable reference sequences can be downloaded from GenBank (http://www.ncbi.nlm.nih.gov/genbank) or the European Nucleotide Archive (http://www.ebi.ac.uk/ena) for many organisms. It is generally impractical to use *breseq* on reference genomes that are >20 Mb in size, and it assumes that the reference genome for clonal samples is haploid.

3. For this tutorial, archives of example input and results files available from the *breseq* website (http://barricklab.org/breseq). Example commands in Subheading 3 use these input read and reference sequence files. The archives also contain examples of output files that illustrate specific points about using *breseq* discussed in Subheading 3.

3　Methods

Commands to be executed in the shell appear in `monospaced font` following a prompt (`$`), which is not part of the typed command. If there is no new prompt at the beginning of a subsequent printed line, then the full command is wrapped to the following line and should be entered into the shell all at once.

3.1 Predicting Mutations in a Clonal Sample

1. In a command-line shell, navigate to inside a directory on your computer system containing both the FASTQ read and GenBank reference files. The commands presented here assume that you are inside the `Clonal_Sample` directory of the downloaded tutorial data. This sample is an Illumina Genome Analyzer NGS dataset of paired-end 36-base reads generated from genomic DNA isolated from a 20,000-generation clone from a long-term laboratory evolution experiment with *E. coli* [6]. The fragment size of the sequenced DNA library was 140 ± 20 bases (s.d.). The reference sequence is the complete genome of the ancestral strain [12]. The example commands presented here can be generalized to any similar re-sequencing analysis of your data.

2. For the example data, the following *breseq* command should be executed from the command prompt and allowed to finish running:

```
$ breseq -j 4 -o Clonal_Output -r REL606.gbk
SRR030257_1.fastq SRR030257_2.fastq
```

Depending on your computer, it may take several hours to complete this *breseq* command. The `-j` option controls how many processes will be used by the pipeline and should be set to the number of CPU cores on your machine for optimal performance. In this example, four cores were used. The `-o` option controls where the analysis files should be directed. In this case, they will be directed to the `Clonal_Output` directory. If this directory does not exist, it will be created. The `-r` option identifies the file `REL606.gbk` as containing the reference genome for comparison to your NGS reads. Finally, all files listed at the end of the line without preceding option flags (`SRR030257_1.fastq` and `SRR030257_2.fastq`) are the read files supplied by the sequencing facility. Typing just `breseq` with no options will display a help message describing other basic options (Fig. 1).

3. It is important to consider what *breseq* settings are appropriate for your data. For the input DNA sequencing reads, paired-end information is not currently utilized (*see* **Note 1**), and 40- to 100-fold coverage of the reference genome is generally optimal for clonal samples (*see* **Note 2**). Reference sequences are expected to be haploid microbial chromosomes or plasmids, and the NGS read dataset for a sample should have nearly complete and even sequencing coverage of each one. Advanced options can relax some of these reference sequence expectations (*see* **Note 3**).

4. After *breseq* has successfully finished executing, several new directories should exist in the `Clonal_Output` path. Directories with numbered prefixes `01`, `02`, `03`, and so on contain intermediate files generated by *breseq* and can typically

```
=========================================================================
breseq 0.24     http://barricklab.org/breseq

Authors: Barrick JE, Borges JJ, Colburn GR, Knoester DB, Meyer AG, Reba A, Strand MD
Contact: jeffrey.e.barrick@gmail.com

breseq is free software; you can redistribute it and/or modify it under the
terms the GNU General Public License as published by the Free Software
Foundation; either version 1, or (at your option) any later version.

Copyright (c) 2008-2010 Michigan State University
Copyright (c) 2011-2013 The University of Texas at Austin
=========================================================================

Usage: breseq -r reference.gbk [-r reference2.gbk ...] reads1.fastq [reads2.fastq ...]
Allowed Options
  -h,--help                          Produce help message showing advanced options
  -o,--output <arg>                  Path to breseq output (DEFAULT=.)
  -r,--reference <arg>               File containing reference sequences in
                                     GenBank, GFF3, or FASTA format. Option may be
                                     provided multiple times for multiple files.
                                     (REQUIRED)
  -n,--name <arg>                    Human-readable name of sample/run for output
                                     [DEFAULT=<none>]
  -j,--num-processors <arg>          Number of processors to use in multithreaded
                                     steps (DEFAULT=1)
  --no-junction-prediction           Do not predict new sequence junctions
  -p,--polymorphism-prediction       Predict polymorphic (mixed) mutations

Utility Command Usage: breseq [command] options ...
  Sequence Utility Commands: CONVERT-FASTQ, CONVERT-REFERENCE, GET-SEQUENCE
  Breseq Post-Run Commands: BAM2ALN, BAM2COV

For help using a utility command, type: breseq [command]

=========================================================================
```

Fig. 1 Basic *breseq* command line help

be ignored and deleted at this point. They allow *breseq* to skip steps that are already complete if its execution is interrupted and restarted. The `output` directory contains files and figures describing mutations predicted in the sample. The `data` directory contains files that can be used to visualize mapped reads and FASTQ files containing reads that did not match the reference genome sequence. If you are executing *breseq* on a remote computer, you will need to copy the `output` and `data` directories back to your computer to view them locally.

5. Open the `summary.html` file located in the `output` directory in a Web browser to bring up a "Summary Statistics" page that displays general information about a *breseq* run (Fig. 2). The "Read File Information" section reports statistics about the reads and how they aligned to the reference sequence. If the percentage of reads aligned is not >90 % in your sample, then you may have a problem with the quality of your sequencing data or it may need further processing before analysis (*see* **Note 4**). Alternatively, the reference genome that you provided may be

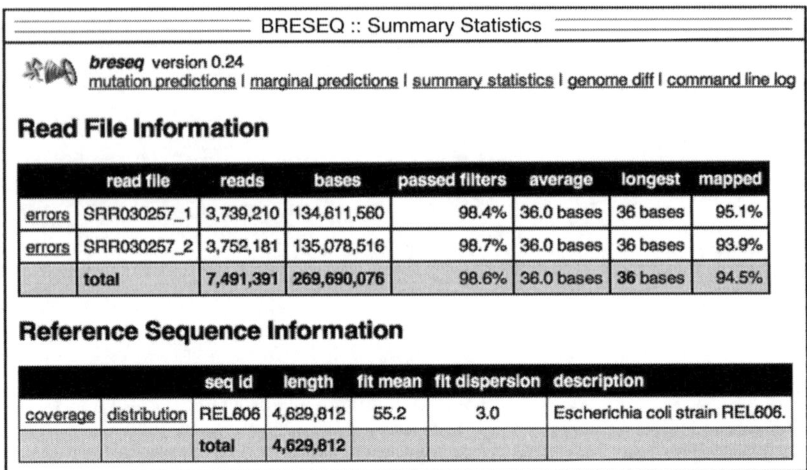

Fig. 2 Example of *breseq* output. The *upper panel* shows a portion of the `summary.html` file which displays general information about the read datasets, reference sequence, and run parameters. The *lower panel* shows part of the main `index.html` page reporting predicted mutations

too divergent from your DNA sample to reliably map reads to it and call mutations (*see* **Note 5**). The "Reference Sequence Information" section reports the depth of read coverage for each reference sequence. The "coverage" links in this table open image files displaying the coverage across each reference sequence that can be used to detect amplifications or deletions of very large genomic regions.

6. Open the `index.html` file located in the `output` directory in a Web browser to display the main "Mutation Predictions" page containing tables listing differences found between the sample and the reference genome (Fig. 2). *breseq* utilizes three types of information to predict mutations: read alignment (RA),

missing coverage (MC), and new junction (JC) evidence. In the main "Predicted mutations" table at the top of this page, links on the left side of this page display alignments and coverage graphs related to the RA, MC, and JC evidence that supported a particular mutation call. The "Unassigned…" tables contain evidence items that also indicate differences between the sample and the reference but that could not be fully resolved to describe precise genetic changes. We discuss how to manually interpret the unassigned evidence in Subheading 3.2. For now, we concentrate on explaining the three types of evidence and how they are used to predict mutations.

7. For reasonable input files, nearly all mutations predicted by *breseq* should be correct. That is, *breseq* should report only evidence for real sequence differences between the sample and the reference. Every single mutation and evidence item predicted from the `Clonal_Sample` data is genuine, for example. However, there is always the possibility of some false-positive predictions passing the statistical tests used by *breseq*, resulting in evidence items appearing in these tables that are not actually well supported upon further examination. As we discuss each type of evidence below, we give advice for recognizing the most common false positives you may encounter. These are further illustrated with *breseq* output files in the `Poor_Evidence_Examples` supplement. The opposite problem of false negatives—where *breseq* does not recover or fully interpret the evidence supporting genuine mutations—is discussed in later sections.

8. RA evidence is derived from analyzing the columns of bases in reads aligned to each position in a reference sequence. *breseq* uses a standard Bayesian SNP caller for RA evidence. The "score" reported is the negative $\log_{10}$ of the posterior probability that the base is not the reported base, corrected for multiple testing by multiplying by the total genome size. In this calculation, *breseq* uses a re-calibrated base error probability model including single-base indels that is fit from the data. RA evidence can support mutations resulting in single-nucleotide substitutions as well as insertions and deletions that are shorter than the read length. Several pieces of adjacent RA evidence may be merged to make a single mutation prediction (e.g., of an insertion of two nucleotides).

9. You can evaluate RA evidence by clicking on the link to bring up a color-coded alignment of reads overlapping the position in question. The most common causes of false-positive predictions can be readily detected on this page. If the base quality scores supporting the variant are uniformly lower (Fig. 3, "RA," center base in right panel) or almost always associated with reads on one strand (Fig. 3, "RA," right panel, mutated "C"

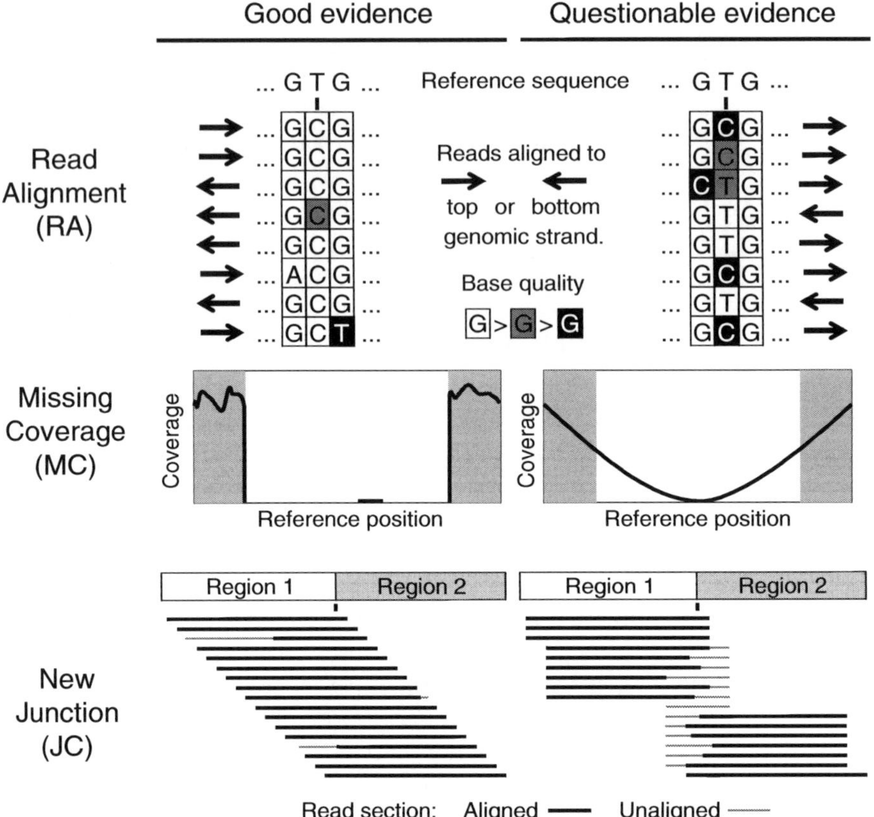

Fig. 3 Evaluating evidence supporting predicted mutations. Characteristics of high-quality (*left column*) and low-quality (*right column*) evidence items that you may encounter in *breseq* output are shown as discussed in the text

base always maps to the top strand indicated by the arrows to the right of the bases), then this may be a problem sequence context for the NGS technology where there is locally an unexpectedly high rate of errors. Alternatively, false-positive RA evidence may originate from mapping reads from the sample back to the reference genome at the wrong site. This may happen when there is a novel sequence in the sample that is not present in the reference, causing reads derived from this new sequence to be mapped incorrectly to their best match in the reference genome. This situation can often be recognized when a subset of the reads supporting a putative RA evidence item do not match across their entire length or have multiple discrepancies from the reference sequence in common (Fig. 4, "Mismapped Reads"). Other false-positive RA evidence may be derived from local misalignment of reads near true examples of short insertions or deletions in the sample. Precautions are taken by *breseq* to not count "masked" bases on each end of a read (shown in lowercase in the alignment) when these could be

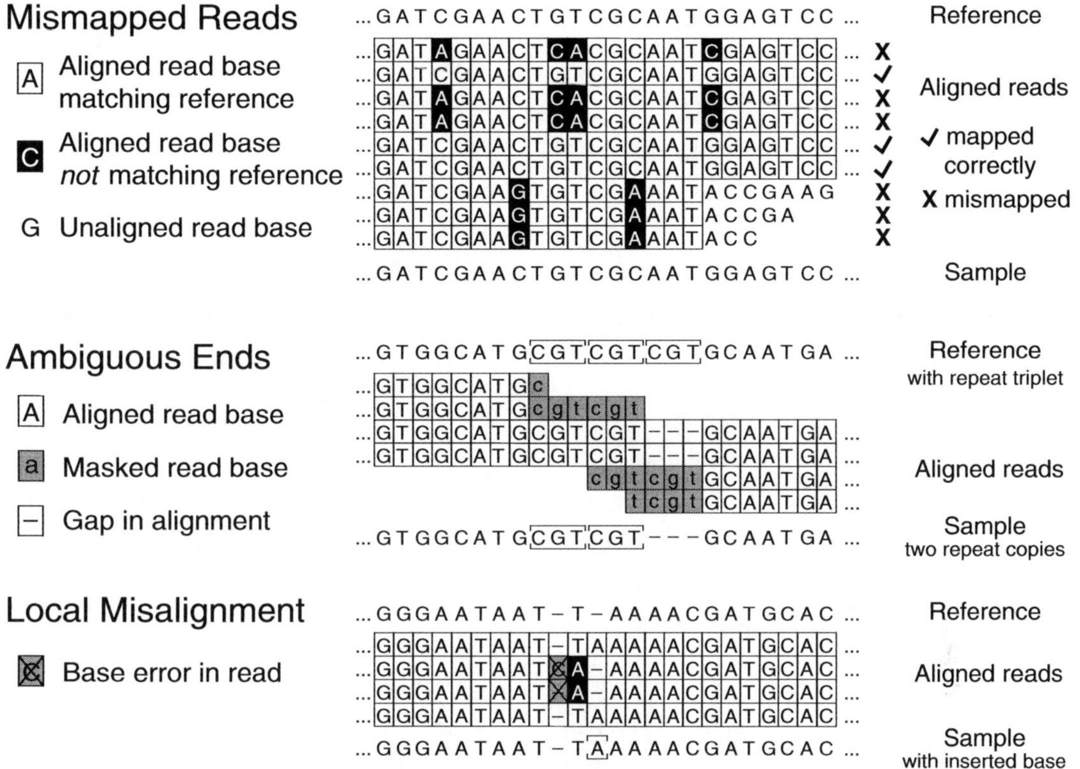

Fig. 4 Possible causes of spurious or low-quality read alignment (RA) evidence. As described in the text, mismapping of reads to an incorrect reference genome site or local misalignment of bases in correctly mapped reads containing base errors can degrade accuracy and sensitivity when predicting micro-indels and single-nucleotide variants

aligned incorrectly if they fell within true expansions or contractions of short sequence repeats (Fig. 4, "Ambiguous Ends"). However, *breseq* does not include a full local read realignment step [4], which may be necessary to make the most optimal prediction in cases where there are base errors in reads in close proximity to true indels or point mutations (Fig. 4, "Local Misalignment").

10. MC evidence is derived from finding places in the genome where no reads align and then extending these intervals in both directions and through repeat regions. Extension of the MC interval is stopped when uniquely mapped read coverage exceeds a threshold that is automatically set by fitting the distribution of read depth coverage found across all normal sites in the current reference sequence. Deletion mutations with precise endpoints are predicted from MC evidence in conjunction with JC evidence or from MC evidence alone when the ends of the deletion are in similarly oriented copies of the same repeat region (Fig. 5). Otherwise, the MC evidence is left unassigned.

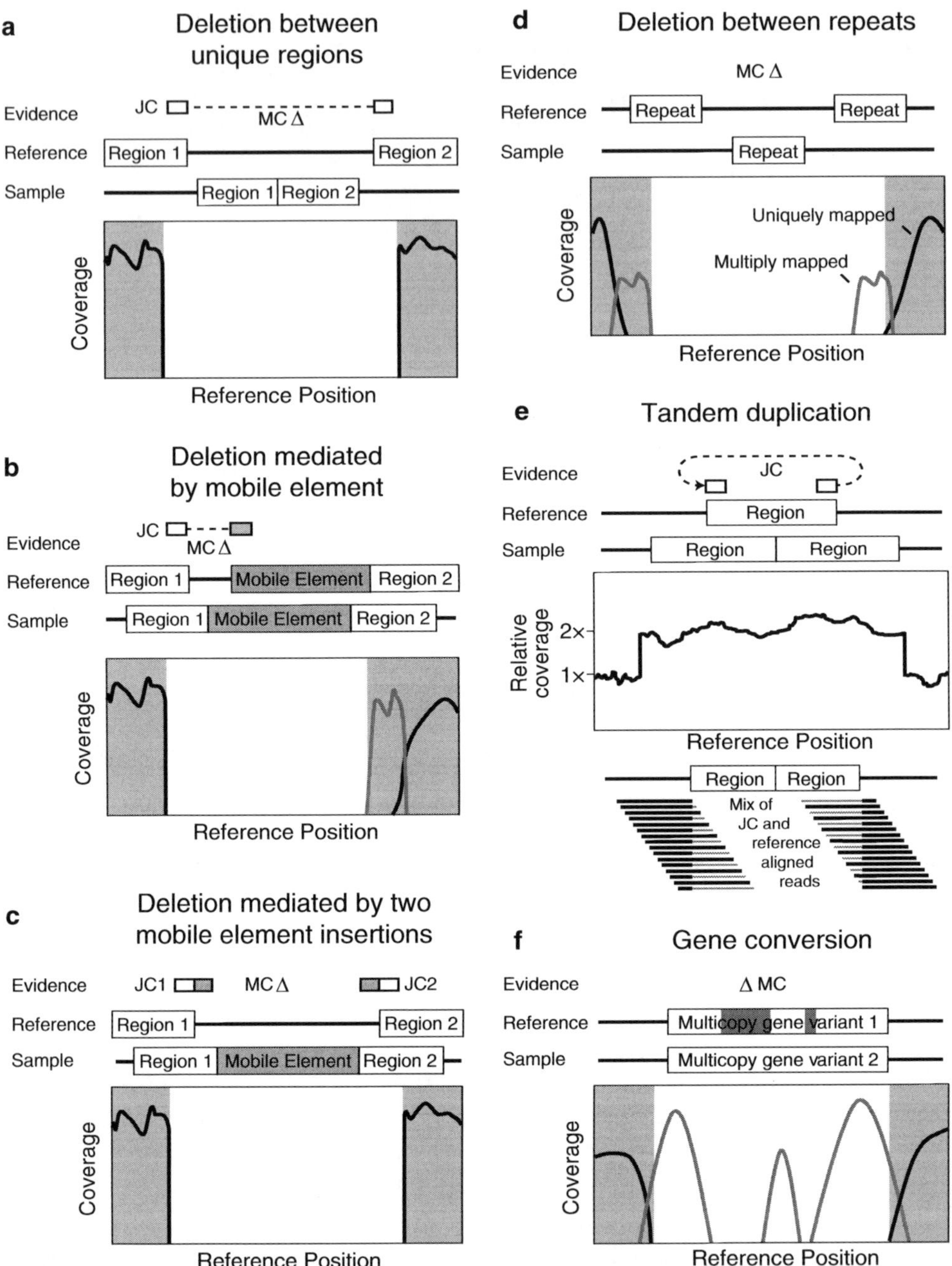

Fig. 5 Evidence supporting complex mutations. In each case schematics of the reference genome and the genome of a sequenced clone are shown. Evidence items that would support the genetic difference between the two genomes are shown above. Relevant graphs of read depth coverage or read alignments are shown below. See the discussion in the text for more details

11. You can evaluate MC evidence by clicking on the link to bring up a graph of read depth coverage at this location. The MC positions have a white background, and the graph is expanded on each end to show the context of surrounding regions with

a gray-shaded background. It is important to understand the difference between "unique" and "repeat" coverage lines in this graph. The former indicates that reads mapped to that location only mapped to one place in the entire reference genome. The latter means reads that mapped there also mapped to other locations in the genome. So even if this example of the repeat was deleted in a sample, you would still find reads that matched this location in the reference. (The contribution to the repeat coverage graph at this location is normalized: if a read mapped four places equally well, it contributes 0.25 to the repeat coverage at each site.) The most common cause of false-positive MC predictions is low sequencing coverage in a sample leading to some regions of the genome not being sampled by chance. This usually gives rise to MC evidence where coverage gradually decreases to zero on each side (Fig. 3, "MC," right panel), rather than the sharp cliffs on each end expected for real deletions in unique regions (Fig. 3, "MC," left panel). You can display the alignment of reads that overlap each edge of the predicted MC evidence through the additional asterisk links shown after selecting it.

12. JC evidence is predicted from split-read matches where read sequences start matching one location in the reference and then "jump" to matching another distant site. *breseq* uses the depth and evenness with which reads are tiled across a putative junction to judge support for it. A bona fide junction should resemble any other site in the genome as far as having an even tiling of reads with many different starting points that map across it (Fig. 3, "JC," left panel). This characteristic is measured in the reported skew, which is the negative $\log_{10}$ probability of the hypothesis that this tiling is unusual. Junction predictions are rejected when they have a high skew. Predicting junctions is computationally expensive and can be disabled using the `--no-junction-prediction` flag if you are not interested in mutations that generate structural variants. JC evidence is used to predict insertions of new copies of mobile genetic elements that create two new junctions between the target site and existing, usually multicopy, sequences corresponding to the element elsewhere in the genome. Note that precise prediction of mobile element insertions requires these elements to be annotated as `repeat_region` features in the input reference files. As mentioned before, JC and MC evidence items are used together to predict deletions with precise endpoints as long as at least one side of the new junction maps to a unique site in the genome (Fig. 5).

13. You can evaluate JC evidence by clicking on the link to bring up an alignment of reads matching across the putative junction better than they match to any position in the reference sequence. False-positive JC predictions can be detected from

uneven tiling of reads across the sequence junction, especially when many reads supporting the junction have unaligned portions at their ends that may indicate that they are being mismapped in the junction context (Fig. 3, "JC," right panel). Additional links from this page display each of the two sites in the original genome connected by the new junction. Usually, these graphs should show mostly one side of a read mapping up to the junction and then the match stopping because the rest of the read supports the new junction and jumps to the distant site in the genome. But in some cases, such as for tandem amplifications, both the old and new junctions may be present, so some reads may align well to the original genome sequence and some may align to the junction. If a side of the JC evidence is highlighted in orange, this indicates that the side falls in a sequence repeat and could also have matched other coordinates in the genome in addition to the one that is displayed. This situation is common for mobile element insertions and for unassigned evidence that requires human intervention to resolve.

14. Of the 45 mutations separating this evolved clone from the ancestor [6], 39 are completely predicted by *breseq* on the first pass. Four of the remaining mutations are not completely predicted by *breseq* because they involve repeat sequences, but they can be resolved from unassigned evidence items, as detailed in the next section. The final two remaining mutations are a large chromosomal inversion mediated by two oppositely oriented copies of an IS*1* transposable element and a base substitution in one of the 28 original copies of IS*1* in the reference genome. They were identified either by Southern blotting [13] or by Sanger sequencing, rather than by analyzing NGS data. Though relatively rare in most samples (2/45 mutations in this one), it is important to be aware of these potential "holes" in a *breseq* analysis. Some types of mutations are difficult or simply impossible to predict from short-read NGS data (*see* **Note 6**).

3.2 Resolving Unassigned Evidence and Mutating a Reference Sequence

1. Using a plain text editor, open the file output.gd located in the output directory of the *breseq* results from running the commands in Subheading 3.1. (Alternatively, download the Clonal_Output archive containing the output and data result directories.) This output file is a "Genome Diff" that describes mutational differences in a sample with respect to the reference genome. The purpose of this *breseq*-specific file is related to, but slightly different from, the Variant Call Format (VCF) file format [14] (*see* **Note 7**). Genome Diffs are tab-delimited text files where each line describes a mutation or a piece of evidence which could support a mutation. A full description of the file format can be found in an appendix of the

documentation included with the *breseq* source code. The `gdtools` command installed as part of *breseq* can be used to perform several types of operations on Genome Diff files (*see* **Note 8**).

2. In addition to false-positive predictions that are possible for the reasons discussed in Subheading 3.1, *breseq* may also sometimes incorrectly rule out a piece of evidence that actually supports a real mutation. These false-negative predictions may sometimes be caught in the "Marginal Predictions" output in the file `marginal.html`, which shows the next best pieces of evidence that fall below score thresholds for further manual inspection. In the example, none of these evidence items support real mutations, but they provide many examples of poor evidence (Fig. 3). False-negative predictions can occur for MC evidence if there is cross-contamination of reads between several samples, some with a deletion and some without it, such that there are enough spurious reads that coverage does not reach zero within a true deletion in a sample.

3. There is a special table titled "Marginal mixed read alignment evidence..." on the "Marginal Predictions" page that reports places in the genome where there is sufficient statistical support for RA evidence supporting that the sample consisted of a mixture of bases at a single reference location, rather than one consensus base. Most of the time (as in this example), these items of evidence are spurious, resulting from poor-quality reads or misalignment, but it is possible that they may represent true mutations. For example, mixed RA evidence with a frequency of around 50 % when sequencing a haploid genome may indicate that a reference region was duplicated and then one of the two copies sustained a point mutation.

4. While on the topic of examining questionable RA evidence, it is important to point out that the files in the `data` directory produced by a *breseq* run can be used to further examine how reads are mapped to the reference genome. This can be accomplished by using the `breseq bam2aln` subcommand, which operates similarly to `bam2cov` (discussed below) but generates an HTML pileup of read alignments for the specified region instead of a coverage graph. Alternatively the `reference.bam`, `reference.fasta` and `reference.gff3` files in the `data` directory can be loaded into Tablet [10] or the Integrative Genomics Viewer [11] to interactively explore how *breseq* mapped reads to the reference genome.

5. In the example, each of the pieces of "Unassigned..." evidence on the main "Mutation Predictions" page supports a real mutation. In the next steps, we explain how to examine the *breseq* output to manually figure out these more complicated types of mutations (Fig. 5). Then, we code them into the

Genome Diff file so that we have a complete manually curated description of all the mutations in the sample. After that, we can apply the Genome Diff to generate a mutated reference sequence and rerun the NGS data against it to verify that we have found (and correctly coded) all of the variation it revealed between the sample and reference genome.

6. For the second unassigned piece of MC evidence that overlaps 23 genes, the endpoints could not be accurately ascertained by *breseq* because they fall in two nearly identical genes (*manB* and *cpsG*). This deletion resulted from a homologous recombination event between equivalent positions in these genes that deleted the intervening sequences and left one hybrid gene copy behind (Fig. 5d). Notice that the ends of this MC evidence item are listed as ranges of positions because they fall in sequence repeats.

7. The `breseq bam2cov` subcommand can generate graphs and tables of coverage over a user-specified region of the genome. This command requires several files as input, but by default it will use the `*.bam` and `*.fasta` files located in a directory named `data` within the current working directory. So, if you run this command from within the main results directory of a *breseq* run you can type

```
$ breseq bam2cov -t -p 0
REL606:2031650-2055600
```

The output `REL606/2031650-2055600.tab` file is tab delimited and can be opened in a spreadsheet program such as `Excel`. The third through sixth columns can be used to determine where the deletion starts and stops based on the breakpoints in the coverage of uniquely mapped reads.

8. When dealing with repeat regions like these, it can also be very helpful to use `BLAST` (on the NCBI Website or locally) to identify all of the other equivalent or near-equivalent copies of the repeat of interest in the genome. In this case, you might `BLAST` the sequence from the left side of the junction (obtained by clicking on the leftmost star of the first MC item):

```
> TGCGCCAGTTTGCTGTTGATCTCACCGCTTGCCG
```

If you query this against the appropriate organism (*Escherichia coli* B str. REL606) on the NCBI Website version of `BLAST`, then you get two perfect matches to coordinates 2031650–2031683 and 2054943–2054976. Note that the deletion can be described in multiple ways to yield the exact same change in the nucleotide sequence of the reference genome, for example, as a deletion of bases 2031650–2054942 or 2031684–2054976.

9. The first MC evidence item and the first JC evidence item describe a deletion between the edge of an existing IS*1* mobile element and a sequence within the ECB_00513 gene.

The coordinates connected by the new junction agree with the ends of the missing coverage and indicate that 8,224 bases corresponding to coordinates 547700–555923 were deleted. This mutation was not fully predicted by *breseq* because there was uncertainty in assigning the ends of the junction. Notice the orange highlighting of the IS*1* side, as expected for a repetitive element, but also that the ECB_00513 side is orange. In fact, the latter junction side maps equally well to the reference starting at position 1604671 and continuing on the reverse genomic strand, explaining why *breseq* did not have the confidence to initially pair these two evidence items. This type of deletion may be mediated by an insertion of a new copy of a mobile element followed by rapid recombination with an existing mobile element copy to delete one copy and the sequence between them (Fig. 5b).

10. The second and fifth pieces of unassigned new junction evidence listed correspond to an IS*150* mobile element insertion into a copy of IS*1* that already exists in the genome. *breseq* does not fully predict this mutation because the insertion could be in any IS*1* copy—the read length is too short to disambiguate which of the 28 ancestral copies—and it arbitrarily predicts junctions to different copies. Annotating this event is therefore best done by looking at the position within the IS*1* feature listed (specifically, 437/768 and 435/768 nt). The true coordinates within one representative copy of IS*1* of each junction end can be confirmed using `BLAST`. In this case, if we mapped both junctions onto the IS*1* element at 241257–242024 in the genome, the first junction would match coordinates upward through 241693 and the other junction would match coordinates 241691 and above. To determine the size of the target site duplication, count how many bases are on both sides of the insertion. Specifically, the 435th, 436th, and 437th bases of the IS*1* (`GTA` it its direction) are now repeated on both sides of the new IS for a target site duplication of three bases. The relative orientation of the inserted IS relative to the genome is determined by identifying how the edges of the IS*150* element map within the IS*1*. In this case, the first base of the IS*150* element (1/1443 nt) is connected to the coordinate 241691 and upward side and the last base of the IS*150* element (1443/1443 nt) is connected to the up-to-coordinate 241693 side, showing that it inserted in the negative orientation within this IS*1* copy (which is itself in the positive genomic direction).

11. The third and fourth unassigned new junctions can also be paired together by running BLAST on each sequence against the REL606 genome. The *ldrC* sequence (`CCGGATAATTCCGGCTT GGTGTGGATACTACTTCTC`) has a single perfect mapping at

1270660–1270692, while the *ldrD* sequence (CCGGTGAGGCGC AATGTTGCGGGGGCTTTATCCCTGG) has five perfect matches including one at positions 1270628–1270663. These two junctions should be paired together based on their proximity to each other (within three bases of one of the alternative best matches of *ldrD*) and that each is paired with the opposite end of an IS*150* insertion. The insertion orientation is determined by identifying what edge of the IS*150* element corresponds to left side of the non-insertion element (up to 1270663 in this case). As in the previous case, this corresponds to the end of the IS*150* element (1977/1977 nt) and the other junction matches the beginning of the element (1/1977 nt) for another negatively oriented insertion. Bases 1270660–1270663 now exist on each side of the newly inserted IS copy, meaning that there was a target site duplication of four bases.

12. In order to verify that our interpretation of the unassigned evidence agrees with the NGS data, we can apply our mutations to the reference genome and re-query the data against the mutated version. The following four lines of text should be added to the `output.gd` file to describe the four mutations that we just manually predicted:

```
DEL_10000_._REL606_547700_8224_mediated=IS1
DEL_10001_._REL606_2031684_23293
MOB_10002_._REL606_241691_IS150_-1_3_ambig-
uous=1
MOB_10003_._REL606_1270660_IS150_-1_4
```

Each item within a line must be separated by tab characters, rather than the underscores shown here for the sake of clarity.

13. Once the additional mutations are added, the `gdtools` `APPLY` command can be run to generate a mutated reference file in GFF3 format:

```
$ gdtools APPLY -r REL606.gbk -f GFF3
-o mutated_reference.gff3
Clonal_Output/output/output.gd
```

14. The new GFF3 file can then be used as a reference file for rerunning *breseq* with the read files to verify that all mutations have been correctly predicted:

```
$ breseq -j 4 -r mutated_reference.gff3
-o mutated_output SRR030257_1.fastq
SRR030257_2.fastq
```

Looking at the resulting `index.html` file should show a "Mutation Predictions" page with no mutations (*see* **Note 9**).

3.3 Additional Examples of Complex Mutations

1. There are several other types of mutations that can be reconstructed from unassigned evidence that you may encounter (Fig. 5). The `Mutation_Examples` supplement contains

breseq output generated from sequencing other clonal samples derived from the same *E. coli* ancestor to illustrate these cases.

2. The first type of mutation is a compound event where there has been a mobile element insertion and then later another mobile element insertion in the same orientation nearby followed by recombination to generate a deletion between them (Fig. 5c). This results in two JC items at the ends to copies of the same mobile element with missing coverage in between. It can be coded in the Genome Diff as two separate events: one mobile element insertion and then a mobile element-mediated deletion. See the `2X_Mobile_Element_&_Deletion` files for an example of this type of mutation, which is more common in longer evolution experiments.

3. The second additional type of mutation is a tandem duplication or amplification (Fig. 5e). This event can be detected as unassigned JC evidence between two unique regions where the original junctions remain present. To confirm that there is the expected increase in coverage of the reference genome region "looped" by this junction and to get a better estimate of its new copy number, use the `breseq bam2cov` subcommand as covered in Subheading 3.2. See the `Amplification` supplementary files for an example of this type of evidence.

4. The third type of mutation that may explain some of the unassigned evidence is a gene conversion event (Fig. 5f). This mutation occurs when a portion of one copy of a near-identical repeat element in a genome is "repaired" to have the same sequence as another repeat copy by recombination. It shows up as MC evidence because reads that used to uniquely map to parts of the repeat that differed by only a few point mutations from the other copies are no longer present. Do not be confused and believe that this missing coverage supports a deletion. The reads that map to this location now map to one of the other copies, so it is true that this sequence is no longer present in the genome. But, it has been replaced at this location with the homologous sequence, rather than deleted. Gene conversions in microbial genomes commonly occur between slightly diverged ribosomal RNA copies. See the supplementary archive `Gene_Conversion` files for an example.

3.4 Identifying and Comparing Variants in Whole-Population Samples

1. Using the shell prompt, navigate to the `Population_Sample` directory in the supplementary archive. The FASTQ files in this directory are for genomic DNA isolated from whole-population samples of *E. coli* after 2,000, 5,000, 10,000, 15,000, and 20,000 generations of a long-term evolution experiment [7]. They consist of 36-base single-end reads generated by an Illumina Genome Analyzer II instrument. The GenBank reference file is the clone that was used to begin the evolution experiment [12]. Genetic diversity changed over

time in this population as new genetic variants arose, competed, and replaced their ancestors. Therefore, each DNA read may be from an individual with a different set of mutations relative to the ancestor (i.e., these samples are "metagenomic").

2. To identify mutations that may be present in only a fraction of a mixed-population sample, add the –p option to the *breseq* command line to switch from consensus mode (the default) to polymorphism prediction mode. The following commands should be run to analyze the population samples for all time points:

```
$ breseq -p -j 4 -o 2K -r REL606.gbk
SRR032370.fastq
$ breseq -p -j 4 -o 5K -r REL606.gbk
SRR032371.fastq
$ breseq -p -j 4 -o 10K -r REL606.gbk
SRR032372.fastq
$ breseq -p -j 4 -o 15K -r REL606.gbk
SRR032373.fastq
$ breseq -p -j 4 -o 20K -r REL606.gbk
SRR032374.fastq
```

In consensus mode, *breseq* assumes that mutations are present in 100 % of a clonal sample. Switching to polymorphism mode enables *breseq* to identify mutations present at intermediate frequencies in the population. Mutations present in 100 % of the mixed sample are still identified, but polymorphic variation is also reported if the evidence for it is statistically significant versus the null hypothesis that the sample was not a mixture. This null hypothesis holds that discrepancies between the reads and the reference genome are adequately explained by sequencing errors.

3. Polymorphism mode output from *breseq* is nearly identical to that for a clonal sample as discussed in Subheading 3.1. The primary addition is a "freq" column, representing the estimated frequency at which each mutation was detected within the population, to the predicted mutations table located on the main `index.html` output page.

4. Predictions of polymorphisms can be bedeviled by various types of non-ideality in the input data where certain sequence contexts or locations are hot spots for sequencing errors at a much greater rate than expected. To reduce the false-positive rate for these predictions *breseq* employs several filters, particularly for polymorphic point mutations. In some cases, it may be desirable to adjust the stringency of these filters (*see* **Note 10**).

5. Commonly, one may want to compare the results of sequencing many related samples side by side to see what mutations are in common between clones or to examine how the frequency of a genetic variant changed over time in a population.

Mutation Comparison

| Predicted mutations | | | | | | | | | |
position	mutation	2K	5K	10K	15K	20K	annotation	gene	description
15,395	G→A					6.0%	intergenic (+99/-48)	*dnaJ* → / → *insL-1*	chaperone Hsp40, co-chaperone with DnaK/IS186 hypothetical protein
16,972	IS*150* (–) +3 bp				86.1%	100%	intergenic (-14/-514)	*mokC* ← / → *nhaA*	regulatory protein for HokC, overlaps CDS of hokC/pH-dependent sodium/proton antiporter
74,220	G→T			14.4%			L22F (TTG→TTT)	*yabI* →	conserved inner membrane protein
161,041	T→G					100%	N302H (AAC→CAC)	*pcnB* ←	poly(A) polymerase I
166,265	C→A			28.4%			A437E (GCA→GAA)	*hrpB* →	predicted ATP-dependent helicase
380,188	A→C			40.7%	92.6%	100%	F239L (TTT→TTG)	*araJ* ←	predicted transporter
430,835	C→T					100%	intergenic (-48/-108)	*insL-2* ← / → *lon*	putative transposase insL for insertion sequence IS186/DNA-binding ATP-dependent protease La
475,288	+G			100%	100%	100%	coding (18/1677 nt)	*ybaL* ←	predicted transporter with NAD(P)-binding Rossmann-fold domain
547,700	Δ8,224 bp			100%	100%	100%	IS1-mediated	[*nmpC*]–[*ECB_00513*]	[nmpC], ybcR, ybcS, ybcT, ybcU, ECB_00510, nohB, ECB_00512, [ECB_00513]
649,391	T→A		100%	100%	100%	100%	I471F (ATC→TTC)	*mrdA* ←	transpeptidase involved in peptidoglycan synthesis (penicillin-binding protein 2)

Fig. 6 Example time course of mutation frequencies in an evolving population. A portion of the comparison file generated from the results of analyzing several whole-population samples is shown. Each column (e.g., 2 K) is for a sample from a different time point (e.g., 2,000 generations)

Once the five *breseq* commands have finished running, the following commands should be run in the `Population_Sample` directory to copy and rename the output Genome Diff files:

```
$ cp 2K/output/output.gd 2K.gd
$ cp 5K/output/output.gd 5K.gd
$ cp 10K/output/output.gd 10K.gd
$ cp 15K/output/output.gd 15K.gd
$ cp 20K/output/output.gd 20K.gd
```

To generate an HTML file with a table comparing all of the mixed-population samples in this example, use the `gdtools` COMPARE subcommand as follows:

```
$ gdtools COMPARE -o compare.html
-r REL606.gbk 2K.gd 5K.gd 10K.gd 15K.gd 20K.gd
```

In the resulting comparison table in file `compare.html`, rows represent specific mutations and columns represent different samples (Fig. 6). In the portion of the table shown, one can see that the *mrdA* mutation arose and was fixed between 2,000 and 5,000 generations. In contrast, it took some time for the later *araJ* mutation to sweep to 100 % frequency.

---

4   Notes

1. For mapping to the reference genome, *breseq* currently treats all input reads as if they are single-end data generated from a DNA fragment library. That is, it does not use paired-end or mate-paired constraints on the orientation and distance between reads. In theory, this information can be used to better predict certain structural variants and mutations in repetitive regions when the distance between read pairs is great enough (*see* **Note 6**). We find that *breseq*'s analysis of just split-read

alignments, where the two halves of one sequencing read match sites that are distant from one another in the reference genome, reliably predicts most new sequence junctions resulting from structural variation. This has also been the conclusion of the developers of other tools such as TopHat-Fusion [5].

2. In our experience, read depth coverage of the reference genome of ≥40-fold gives very accurate mutation predictions for Illumina data. Greater coverage beyond this usually does not result in any improvements and can make the pipeline take substantially longer to run. So, it may be desirable to truncate overly large FASTQ input files to an estimated ~100-fold coverage of the reference genome to decrease runtime in some cases. This can be easily accomplished by using Unix commands like `head` to extract a subset of lines from a large FASTQ file. The exception to this rule is for whole-population samples (Subheading 3.4) where read depth can limit the discovery of rare variants and as much read data as possible should generally be used.

3. If you performed some sort of enrichment for DNA from specific genomic regions in preparing your sample (e.g., you are sequencing PCR amplicons), add the `-t` flag to the *breseq* command to activate targeted sequencing mode. This relaxes the assumption that there will be equal coverage over the entire reference sequence and prevents analysis steps that only apply when this is true (e.g., predicting deletions from missing coverage). If your sample has foreign DNA in it (such as transposons, viruses, or plasmids) and your only interest in mutations involving these sequences is in their potential insertion into the host chromosome (for example, you sequenced a transposon insertion mutant library and have the sequence of a suicide plasmid that carried the transposon), reference files for each of the foreign genomes should be supplied using the junction-only reference (`-s`) option in addition to the main reference of interest which is supplied using the typical `-r` option. This usage provides three main benefits: SNPs, indels, and other mutations affecting only the foreign DNA are ignored, insertions into the genome of interest will be correctly identified (e.g., where the transposon from the suicide plasmid integrated), and it prevents reads originating from the foreign DNA from accidentally mismapping to the genome of interest where they could possibly lead to spurious mutation calls.

4. It is not unusual for 5–10 % of total reads in any NGS dataset to not align to the reference genome because they pass quality filters but contain many incorrect base calls. A higher percentage of reads not mapping may indicate a poorly constructed DNA fragment library containing many adapter dimers, for example. Another common problem is not removing barcode

sequences at the ends of reads after de-multiplexing samples. This can result in suboptimal mapping of reads and spurious mutation predictions in *breseq*. We suggest checking the quality of your input FASTQ file using a tool such as FastQC that can detect these and other concerns [15]. Tools like FLEXBAR [16] can be used to trim adaptor and bar-code sequences before they are used as input for *breseq*.

5. The re-sequencing strategy employed by *breseq* breaks down when there is sufficient sequence divergence between the sample and the reference genome to lead to mismapping of reads, such as when there is a high level of local nucleotide divergence or there are novel sequences in the sample that are not in the reference genome. In these cases, it may be valuable to use de novo assembly tools, such as Velvet [17] or ALLPATHS-LG [18], on either the unmapped reads after a *breseq* run (located at `data/unmatched.*`) or all reads and then compare assemblies to the reference genome or to one another using tools for comparing whole-genome sequences, such as MUMmer [19].

6. Due to inherent limitations in the information present in short read data and the algorithms used by *breseq*, certain kinds of mutations will never be predicted. These are generally related to sequence repeats in the reference genome that are longer than the read length, such as multi-copy mobile genetic elements, ribosomal RNA operons, and recently duplicated genes. It may be impossible to span these repeats with a single read to anchor the two unique sides relative to each other. This may result in an inability to detect the new junctions formed by large chromosomal inversions, deletions, or tandem amplifications that occur through equivalent sites of the same repeat. Deletions between these elements can be found by looking for missing coverage (as detailed for the *manB–cpsG* deletion). Duplications or amplifications should result in higher read coverage of the amplified region between repeats. This type of copy number variation is not currently automatically predicted by *breseq*, but it can be detected by generating and manually examining read coverage graphs that tile the genome at a high enough resolution (see options for `breseq bam2cov`). Chromosomal inversions through these repeats will not result in a change in coverage and therefore cannot be predicted. An inability to map reads uniquely to multi-copy sequence repeats can also make it difficult to detect a new point mutation in any one copy of the repeat element. Theoretically, these mutations could be detected by looking for polymorphisms within the "population" of the repeat sequence within a single clonal genome, but *breseq* does not currently attempt this complicated analysis.

7. Files in the community VCF format describe sequence variation between a sample and a reference genome [14]. Genome Diff files output by *breseq* list mutational events separating two genomes. This is a subtle but important difference. Currently, Genome Diff files can encode additional information regarding evolutionary processes and mechanisms that does not fit naturally in a VCF file. For example, imagine a complex mutation that resulted from a new copy of a mobile element inserting in a genome and then later recombination between this element and an existing copy of the mobile element resulting in a large deletion. A VCF file describing this sequence variation would record the large deletion only. A Genome Diff file could include entries for each of the two separate mutations, so that they could be counted separately or used to construct a better phylogenetic tree relating multiple samples. Another example where information would be lost in a VCF file is when a point mutation occurs within a region that is later deleted. It is possible that future versions of VCF will have ways to precisely describe these situations. For now, Genome Diff files can be converted to standard VCF files for display in NGS browsers such as the Integrated Genomic Viewer (IGV) using the `gdtools GD2VCF` subcommand.

8. In addition to the subcommands described in the text for applying and comparing Genome Diff files, `gdtools` provides basic support for set operations (`UNION`, `INTERSECTION`, `SUBTRACT`), format conversion (`GD2VCF`), analyzing the overall characteristics of mutations (`COUNT`), or drawing images of genomes showing mutations (`GD2CIRCOS`). Be aware that commands that rely on outside software or formats are not always up to date or as stable as core *breseq* operations.

9. If any evidence has been discarded as a false positive, it will likely still be listed again after the *breseq* run on the mutated genome. Any new mutations or evidence found as a result of re-querying the NGS data against the updated reference will have their locations relative to the mutated genome and not to the original genome. So, be sure that any additional updates to the original Genome Diff file that you make are based on the original genome coordinates and not the updated genome.

10. *breseq* employs several filters to attempt to catch the bulk of false-positive predictions of polymorphisms that are not ruled out by the underlying error model [7]. First, many spurious polymorphism predictions can be recognized because all of the reads supporting the variant correspond to only one of the two genomic strands, rather than occurring evenly on both strands as would be expected for a real mutation. *breseq* uses Fisher's exact test to judge the significance of this bias. Second, the bases supporting a variant may have uniformly lower quality

scores than those supporting the consensus. *breseq* uses a Kolmogorov–Smirnov test to detect this bias. The `--poly-morphism-bias-cutoff` option sets the p-value cutoff for both of these tests. Lower values of this parameter will reject fewer polymorphism predictions. The option `--polymor-phism-minimum-coverage-each-strand` can also be used to add a hard requirement that a certain number of reads on each strand must support a variant for it to be reported. Indels are rare in Illumina data and very common in Roche 454 data. Each situation can lead to overestimating the significance of this type of variation, so there are options to specifically not predict polymorphisms in repeats of a single base that exceed a certain length (`--polymorphism-reject-homo-polymer-length`) or to not predict any indel polymorphisms (`--polymorphism-no-indels`). Finally, polymorphisms with lower frequencies are more prone to misprediction, so you can apply a simple frequency cutoff criterion to all predictions using the `--polymorphism-frequency-cutoff` option. Predicting polymorphisms is inherently noisier than predicting consensus mutations. You may need to optimize these parameters for characteristics of your particular samples to achieve the best results.

Acknowledgements

D.E.D. was supported by a University of Texas at Austin CPRIT Cancer Research Traineeship. Development of *breseq* has been supported by an NSF Postdoctoral Research Fellowship in Biological Informatics (DBI-0630687) and by grants from the NSF BEACON Center for the Study of Evolution in Action (DBI-0939454), NIH (R00-GM087550), and CPRIT (RP130124) to J.E.B. Additional programmers and users who have provided valuable feedback and bug reports are thanked in the *breseq* documentation.

References

1. Mardis ER (2008) Next-generation DNA sequencing methods. Annu Rev Genomics Hum Genet 9:387–402

2. Eid J, Fehr A, Gray J et al (2009) Real-time DNA sequencing from single polymerase molecules. Science 323:133–138

3. Trapnell C, Salzberg SL (2009) How to map billions of short reads onto genomes. Nat Biotechnol 27:455–457

4. DePristo MA, Banks E, Poplin R et al (2011) A framework for variation discovery and genotyping using next-generation DNA sequencing data. Nat Genet 43:491–498

5. Kim D, Salzberg SL (2011) TopHat-fusion: an algorithm for discovery of novel fusion transcripts. Genome Biol 12:R72

6. Barrick JE, Yu DS, Yoon SH et al (2009) Genome evolution and adaptation in a long-term experiment with *Escherichia coli*. Nature 461:1243–1247

7. Barrick JE, Lenski RE (2009) Genome-wide mutational diversity in an evolving population of *Escherichia coli*. Cold Spring Harb Symp Quant Biol 74:119–129

8. Woods RJ, Barrick JE, Cooper TF et al (2011) Second-order selection for evolvability in a

large *Escherichia coli* population. Science 331: 1433–1436

9. Blount ZD, Barrick JE, Davidson CJ, Lenski RE (2012) Genomic analysis of a key innovation in an experimental *Escherichia coli* population. Nature 489:513–518

10. Milne I, Stephen G, Bayer M et al (2013) Using Tablet for visual exploration of second-generation sequencing data. Brief Bioinform 14:193–202. doi:10.1093/bib/bbs012

11. Thorvaldsdóttir H, Robinson JT, Mesirov JP (2013) Integrative Genomics Viewer (IGV): high-performance genomics data visualization and exploration. Brief Bioinform 14:178–192

12. Jeong H, Barbe V, Lee CH et al (2009) Genome sequences of *Escherichia coli* B strains REL606 and BL21(DE3). J Mol Biol 394: 644–652

13. Schneider D, Duperchy E, Coursange E et al (2000) Long-term experimental evolution in *Escherichia coli*. IX. Characterization of insertion sequence-mediated mutations and rearrangements. Genetics 156:477–488

14. Danecek P, Auton A, Abecasis G et al (2011) The variant call format and VCF tools. Bioinformatics 27:2156–2158

15. Andrews S FastQC: a quality control tool for high throughput sequence data. http://www.bioinformatics.babraham.ac.uk/projects/fastqc/

16. Dodt M, Roehr J, Ahmed R, Dieterich C (2012) FLEXBAR—flexible barcode and adapter processing for next-generation sequencing platforms. Biology 1:895–905

17. Zerbino DR, Birney E (2008) Velvet: algorithms for de novo short read assembly using de Bruijn graphs. Genome Res 18:821–829

18. Ribeiro FJ, Przybylski D, Yin S et al (2012) Finished bacterial genomes from shotgun sequence data. Genome Res 22:2270–2277

19. Kurtz S, Phillippy A, Delcher AL et al (2004) Versatile and open software for comparing large genomes. Genome Biol 5:R12

Chapter 13

3D-Fluorescence In Situ Hybridization of Intact, Anaerobic Biofilm

Kristen A. Brileya, Laura B. Camilleri, and Matthew W. Fields

Abstract

FISH (fluorescence in situ hybridization) is a valuable technique to visualize and quantify localization of different microbial species within biofilms. Biofilm conformation can be altered during typical sample preparation for FISH, which can impact observations in multispecies biofilms, including the relative positions of cells. Here, we describe methods to preserve 3-D structure during FISH for visualization of an anaerobic coculture biofilm of *Desulfovibrio vulgaris* Hildenborough and *Methanococcus maripaludis*.

Key words Biofilm structure, Multispecies biofilm, Population interactions, Sulfate-reducing bacteria, Methanogenic archaea

1   Introduction

FISH techniques based upon rRNA gene sequences have revolutionized the study of microorganisms in natural assemblages and environments and have allowed scientists and engineers to characterize the structure of microbial communities more quantitatively [1]. Techniques continue to improve in terms of sensitivity [2] and in situ hybridization can be used in combination with other methods [3] to elucidate information about metabolic activity, genetic potential, and environmental conditions [4–7]. Structure and local environments are particularly important in the study of microbial biofilms, and the application of FISH techniques to microbial assemblages provides insight into the complex structure–function relationships of microbial biofilms and/or aggregates.

Biofilms are typically defined as self-assembled groups of cells adhered to surfaces that are embedded within an exopolymer matrix. Matrices of well-studied biofilms are usually composed of exopolysaccharides (EPS) that can include carbohydrate, protein, DNA, and various appendages (i.e., pili, fimbriae) within the matrix [8]. As FISH techniques that can probe complex multispecies

Lianhong Sun and Wenying Shou (eds.), *Engineering and Analyzing Multicellular Systems: Methods and Protocols,*
Methods in Molecular Biology, vol. 1151, DOI 10.1007/978-1-4939-0554-6_13, © Springer Science+Business Media New York 2014

biofilms continue to emerge (e.g., ref. [9]), the maintenance of biofilm structure and integrity will be crucial to deciphering structure–function relationships. Here, we demonstrate the use of 3D-FISH on an anaerobic, coculture biofilm using the techniques previously described by Daims et al. [10].

2   Materials

2.1   Cover Slip Preparation

1. Absolute ethanol.

2. Acidic ethanol: Prepare fresh 99 mL of 70 % ethanol in a small beaker by adding 70 mL of absolute ethanol to 29 mL deionized water. Add 1 mL of 12 M HCl for a final concentration of 1 % (v/v).

3. Bind-Silane working solution: Add 1 mL of Bind-Silane (Amersham Biosciences, Uppsala, Sweden), 3 mL of 10 % (v/v) glacial acetic acid, and 296 mL of deionized water. Mix until the solution is clear. Store at 4 °C.

4. Cover slips (No. 1½, 24 × 50 mm).

2.2   Sample Preparation, Fixation, and Storage

1. 3× Phosphate-buffered saline (PBS): Prepare each of A. 200 mM NaH_2PO_4, B. 200 mM Na_2HPO_4, and C. 390 mM NaCl then adjust the pH of solution B to 7.2–7.4 with solution A. Next add 150 mL of the prepared phosphate buffer to 850 mL of solution C, and adjust the final pH to 7.2–7.4 with NaOH. To make 1× PBS dilute this solution with two volumes of deionized water.

2. 4 % Paraformaldehyde (PFA): Heat 33 mL of deionized water to 65 °C and add 2 g of paraformaldehyde while stirring (*see* **Note 1**). Add NaOH until the paraformaldehyde dissolves, then add 16.6 mL 3× PBS. Cool to room temperature and adjust pH to 7.2–7.4 then filter with a 0.2 μm filter. Store at –20 °C.

3. Glycerol.

4. Ice.

2.3   Polyacrylamide Embedding

1. Polyacrylamide (PAA) solution: Prepare 20 % (w/v) solution in water or 1× PBS. A 37.5:1 solution of Acrylamide: BIS Acrylamide (EMD Chemicals, Inc., Darmstadt, Germany) (*see* **Note 2**) is 40 % (w/v), so the 20 % PAA solution is prepared with an equal volume of water or 1× PBS. Store at 4 °C.

2. Ammonium persulfate (APS): A 10 % (w/v) solution is made by adding 1 g of APS (*see* **Note 3**) to 10 mL of deionized water. Store at 4 °C.

3. TEMED (Tetramethylethylenediamine) (*see* **Note 4**).

4. Polytetrafluoroethylene (PTFE) sheets: PTFE or Teflon™ (McMaster Carr, USA) can be obtained in a variety of dimensions and preparations. Use either polished "adhesive-ready" PTFE or a type that already has adhesive. Choose a thickness that is as tall as or taller than your biofilm, that your microscope objective has a long enough working distance to see through. For this study a 0.01 in. thick (~254 μm) film with adhesive was used. Cut small frames that fit around the perimeter of your cover slips.

2.4 In Situ Hybridization

1. 50, 80, 96 % (v/v) ethanol.

2. 5 M NaCl: Prepare 50 mL by dissolving 14.61 g of NaCl in water. Store at room temperature.

3. 1 M Tris–HCl buffer prepared in water, adjust pH with 10 N NaOH to 8.0. Store at room temperature.

4. High-quality molecular grade deionized formamide (*see* **Note 5**). Store at 4 °C or according to manufacturer's recommendation. Low-quality formamide that is contaminated with cations will reduce hybridization stringency.

5. Sodium dodecyl sulfate (SDS): Prepare a 10 % (w/v) solution by dissolving 2 g of SDS in 20 mL of deionized water. Store at room temperature.

6. 0.5 M Ethylenediaminetetraacetate (EDTA sodium salt) prepared in water, adjust pH with NaOH pellets to 8.0. As pH approaches 8.0, EDTA will dissolve slowly. At this point, use 10 N NaOH to raise pH with more control. Store at room temperature.

7. Fluorescently labeled oligonucleotide probe at 30 ng/μL for most fluorophores, or 50 ng/μL for probes labeled with fluorescein and its derivatives. For this example (Fig. 1), domain-level probes EUB338 (5′- GCT GCC TCC CGT AGG AGT -3′) double-labeled with Cy3 (5′and 3′) and ARCH915 (5′- GTG CTC CCC CGC CAA TTC CT -3′) double-labeled with Cy5 (5′and 3′) were used (Thermo Scientific Custom Biopolymers, Ulm, Germany). Store at –20 °C in the dark.

8. Ice-cold deionized water.

9. Citifluor AF1 Antifadent (Citifluor Ltd., London, UK).

10. Hybridization oven at 46 °C.

11. Water bath at 48 °C.

12. Oil free compressed air.

13. Microscope equipped with appropriate filter sets for selected fluorophore-labeled probe. A confocal laser scanning microscope (CLSM) is necessary to acquire optical sections through the depth of the biofilm. A long working distance objective is

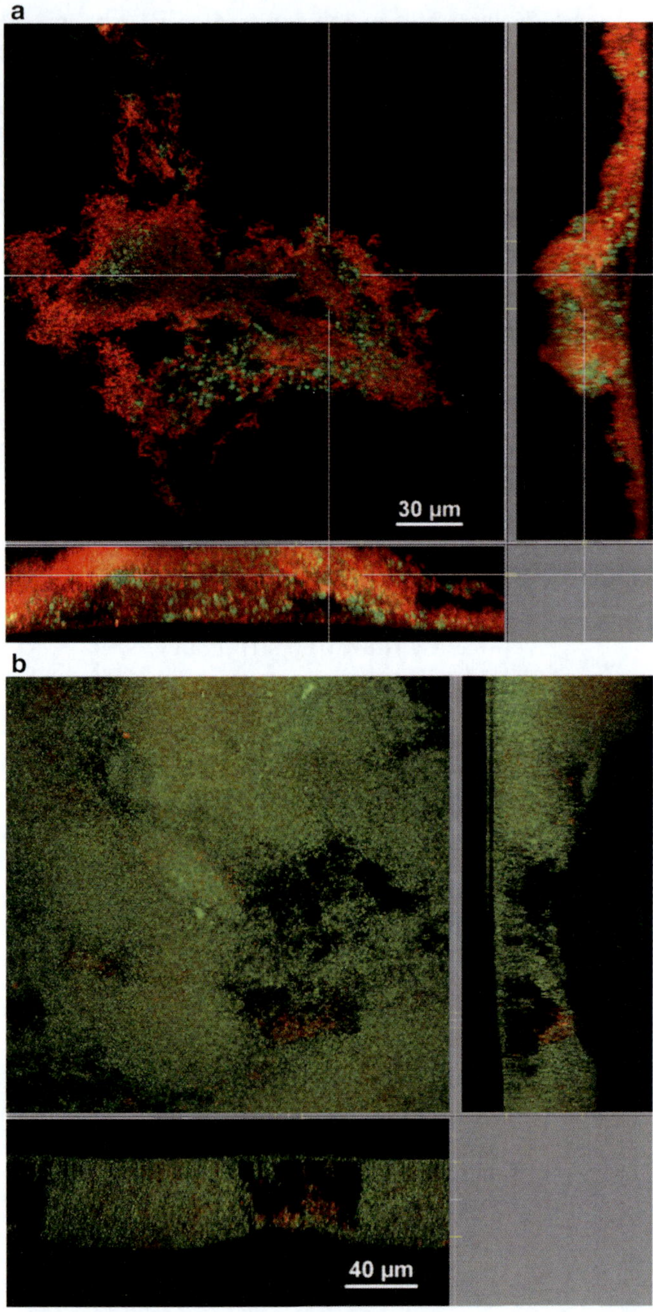

Fig. 1 (**a**) Coculture anaerobic biofilm embedded in polyacrylamide and hybridized to probes Arch915 (*green*) for *M. maripaludis* and Eub338 (*red*) for *D. vulgaris*. *Main panel* is biofilm as viewed from above, while *side panels* represent vertical cross sections. (**b**) *D. vulgaris* (Arch915—*Red*) and *M. maripaludis* (Eub338—*Green*) coculture biofilm embedded in polyacrylamide and viewed from above. Images were taken on a Leica TCS SP5 II upright confocal laser scanning microscope using a 63× long working distance water dipping objective. Digital reconstructions were done using Imaris v.7.6 (Bitplane, Zurich, CH). From colocalization analysis of these images using MetaMorph v. 7.6 (Molecular Devices, Sunnyvale, CA), we observed that the populations were distributed randomly through the depth of the biofilm

also required to obtain optical sections through the gel pad, such as the Leica 63× HCX APO L U-V-I water dipping objective (#506148, Leica Microsystems, Inc., Germany) used in the described example.

---

3   Methods

3.1   Cover Slip Pretreatment

1. Clean cover slips by dipping in a small beaker of acidic ethanol and drying on a lint-free tissue. To coat cover slips so that later they can be covalently attached to biofilm-embedded polyacrylamide gel, submerge the cover slips in the Bind-Silane solution and incubate at room temperature for 1 h. Rinse the cover slips in deionized water and then again in ethanol. Dry the cover slips on a lint-free tissue. Cover slips can be stored for several months at room temperature.

3.2   Sample Preparation, Fixation, and Storage

1. Whole biofilm grown on the microscope slide should be fixed immediately while still hydrated by placing in a 50 mL conical tube or slide holder with ice-cold 4 % PFA. Aqueous samples are typically fixed in three volumes of 4 % PFA to one volume of 1× PBS, especially if the biofilm has a low moisture content. A wet biofilm may be fixed directly in 4 % PFA. In both cases, make sure that the slide is submerged in the fixative solution. Fixation is typically for 3–12 h, and longer fixation times may render cells impermeable to the oligonucleotide probe (*see* **Note 6**).

2. Following fixation, slides should be stored at –20 °C in 1:1 glycerol:1× PBS. Do not store in ethanol as this will inhibit the polymerization of acrylamide during embedding.

3.3   Polyacrylamide Embedding

All of these steps must be done very quickly, so assemble all materials first in the fume hood. Leave the biofilm slide in the PBS:glycerol solution until you are ready to use it.

1. Lay pretreated cover slip on a paper towel and apply one of the PTFE frames to the edge by pressing gently with forceps.

2. Prepare a working PAA solution with 50 µL of 20 % PAA, 0.5 µL of 10 % APS, and 0.5 µL of TEMED. Immediately pipet this solution onto the biofilm and place the pretreated cover slip on the biofilm with the Teflon frame facing the solution and biofilm (Fig. 2). Make sure that there are no air bubbles as this will prevent the polymerization of acrylamide, then weight the slide down very lightly with a small serum bottle. Let the acrylamide solution polymerize for 10–15 min at room temperature. A larger volume may be necessary to cover the entire biofilm.

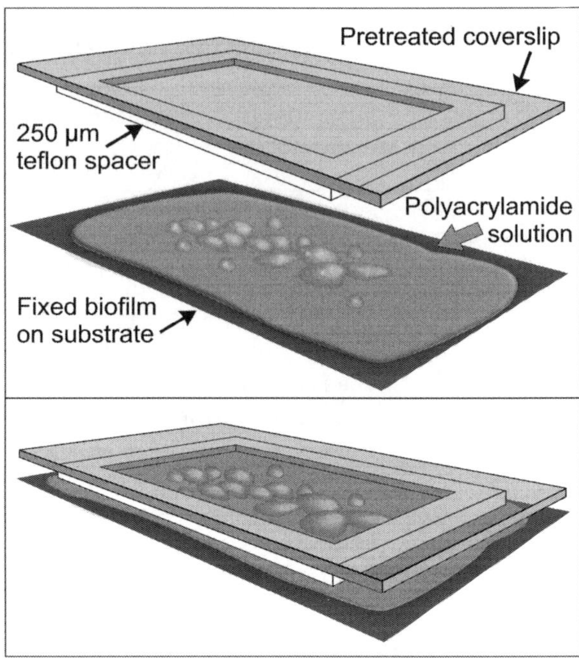

Fig. 2 Schematic showing assembly of the cover slip with Teflon frame and biofilm embedded in polyacrylamide

3. When the polyacrylamide has solidified to a gel, carefully separate the slide from the cover slip. The biofilm should now be upside down on the cover slip. To ensure that all of the biofilm came off the slide, you can examine it with light or fluorescent microscopy using an appropriate stain (*see* **Note** 7).

3.4 In Situ Hybridization

1. Dehydrate the whole cover slip and gel pad in an increasing ethanol series (5 min each in 50, 80, and 96 % ethanol). Dry gel pad completely by blowing oil-free compressed air on the surface for approximately 30 s. It is normal for the gel pad to turn white during dehydration.

2. Thaw oligonucleotide probes on ice in the dark and prepare hybridization buffer. Prepare hybridization buffer in a 1.5 mL centrifuge tube containing 5 M NaCl, 1 M Tris, deionized water, formamide, and 10 % SDS in the concentration appropriate for the probe as in selected column of Table 1. The appropriate formamide concentration is reported in the literature for previously published probes, and must be determined experimentally when a new probe is developed (e.g., ref. [11]).

3. Add 200 μL of hybridization buffer and 4 μL of probe to the surface of the gel pad and mix up and down by pipetting. The hybridization buffer will dome up on the surface of the gel for a short time, and it will not be difficult to mix in the probe,

Table 1
Volume in μL of each component of hybridization buffer to be used at each formamide concentration from 0 to 70 %

Form %	0 %	5 %	10 %	20 %	25 %	30 %	35 %	40 %	50 %	55 %	70 %
NaCl	180	180	180	180	180	180	180	180	180	180	180
Tris	20	20	20	20	20	20	20	20	20	20	20
H_2O	800	750	700	600	550	500	450	400	300	250	100
Form	0	50	100	200	250	300	350	400	500	550	700
SDS	1	1	1	1	1	1	1	1	1	1	1

Table 2
Volume of each component of washing buffer to be used at the corresponding formamide concentration from 0 to 70 %

Form %	0 %	5 %	10 %	20 %	25 %	30 %	35 %	40 %	50 %	55 %	70 %
NaCl	9 ml	6.3 ml	4.5 ml	2,150 μL	1,490 μL	1,020 μL	700 μL	460 μL	180 μL	100 μL	0
Tris (ml)	1	1	1	1	1	1	1	1	1	1	1
EDTA (μL)	0	0	0	500	500	500	500	500	500	500	500
H_2O (ml)	fill to 50	fill to 50	fill to 50	fill to 50	fill to 50	fill to 50	fill to 50	fill to 50	fill to 50	fill to 50	fill to 50
(SDS) (μL)	50	50	50	50	50	50	50	50	50	50	50

however take care to avoid touching the gel pad with the pipet tip. Work quickly to avoid exposing the probes to light and to prevent the hybridization buffer from evaporating.

4. Prepare a humid chamber hybridization tube (50 mL conical tube) by folding a Kimwipe tissue and laying it in the tube. Pour the rest of the hybridization buffer onto the tissue to keep the atmosphere in the chamber humid and at the desired concentration of formamide. Immediately transfer the cover slip into the hybridization tube, laying it on the tissue with the gel pad facing up, and incubate horizontally in the hybridization oven (46 °C) for 2–3 h. Longer hybridization times may be required for very thick or dense biofilms.

5. Prepare washing buffer in a 50 mL conical tube for the appropriate stringency according to one column of Table 2. Use the same concentration column as chosen for the hybridization

buffer, based on the reported value for the probe used, or the value determined for a newly developed probe. Warm the washing buffer in a water bath (48 °C).

6. After hybridization is complete, carefully and quickly transfer the embedded sample into the preheated washing buffer and incubate vertically for 35–40 min in the water bath (48 °C). The transfer must be done quickly and in the fume hood to avoid inhaling warm formamide and to prevent the wash buffer from cooling.

7. Dip the cover slip quickly in and out of ice-cold deionized water eight times to cool and wash the polyacrylamide pad and biofilm and dry immediately using oil-free compressed air.

8. Place the cover slip in the dark at room temperature for an additional 10 min to ensure it is dried completely. Do not over dry as the gel pad will crack.

9. Image the PAA embedded sample using a long working distance objective on a CLSM. Imaging can be done through the cover slip or through the gel pad.

4 Notes

1. Paraformaldehyde is a suspected carcinogen, wear gloves and a dust mask and work in a fume hood.

2. Acrylamide is known to cause cancer and birth defects. Wear gloves and work in a fume hood.

3. APS is a strong oxidant and irritant.

4. TEMED is highly flammable and corrosive.

5. Formamide is a known teratogen and causes irritation upon contact or inhalation. May cause unconsciousness. It is especially volatile when warm and should be used in the fume hood and handled with care.

6. Fixation time and requirements vary between cell types (e.g., Gram-negative, Gram-positive, or Archaea with or without an S layer) and should be tested on aqueous cultures first whenever possible. Fixation with one volume of 96 % ethanol mixed with one volume of 1× PBS may be preferable in some cases, and the fixed cells can be immediately stored at –20 °C. For some Gram-positive organisms, it may be necessary to treat with enzymes such as lysozyme, proteinase k, or achromopeptidase to permeabilize the cell.

7. To practice preparing the polyacrylamide gel pad and test the capabilities of your microscope, you can first try embedding and imaging fluorescent beads, for example, Constellation Microspheres (Invitrogen Molecular Probes, USA).

Acknowledgements

The authors wish to thank Betsey Pitts for microscopy assistance and Dr. Sebastian Lücker for thoughtful discussions. Special thanks to Peg Dirckx for preparing Figs. 1 and 2. This work was supported as a component of ENIGMA, a scientific focus area program supported by the U.S. Department of Energy, Office of Science, Office of Biological and Environmental Research, Genomics: GTL Foundational Science through contract DE-AC02-05CH11231 between Lawrence Berkeley National Laboratory and the U.S. Department of Energy. K.A.B. and L.B.C. were also supported by a NSF-IGERT fellowship in Geobiological Systems at Montana State University (DGE 0654336). The confocal microscopy equipment used was purchased with funding from the NSF-Major Research Instrumentation Program and the M.J. Murdock Charitable Trust.

References

1. Wagner M, Haider S (2012) New trends in fluorescence *in situ* hybridization for identification and functional analyses of microbes. Curr Opin Biotechnol 23:96–102

2. Pernthaler A, Pernthaler J, Amann R (2002) Fluoresecence in situ hybridization and catalyzed reporter deposition for the identification of marine bacteria. Appl Environ Microbiol 68:3094–3101

3. Wagner M (2009) Single-cell ecophysiology of microbes as revealed by Raman microspectroscopy or secondary ion mass spectrometry imaging. Annu Rev Microbiol 63:411–429

4. Dekas AE, Poretsky RS, Orphan VJ (2009) Deep-sea Archaea fix and share nitrogen in methane-consuming microbial consortia. Science 326:422–426

5. Gieseke A, Purkhold U, Wagner M, Amann R, Schramm A (2001) Community structure and activity dynamics of nitrifying bacteria in a phosphate-removing biofilm. Appl Environ Microbiol 67:1351–1362

6. Huang WE, Stoecker K, Griffiths R, Newbold L, Daims H, Whiteley AS, Wagner M (2007) Raman-FISH: combining stable-isotope Raman spectroscopy and fluorescence in situ hybridization for the single cell analysis of identity and function. Environ Microbiol 9: 1878–181889

7. Wagner M, Nielsen PH, Loy A, Nielsen JL, Daims H (2006) Linking microbial community structure with function: fluorescence in situ hybridization-microautoradiography and isotope arrays. Curr Opin Biotechnol 17:1–9

8. Branda SS, Vik A, Friedman L, Kolter R (2005) Biofilms: the matrix revisited. Trends Microbiol 13:20–26

9. Valm AM, Mark Welch JL, Rieken CW, Hasegawa Y, Sogin ML, Oldenbourg R, Dewhirst FE, Borisy V (2011) Systems-level analysis of microbial community organization through combinatorial labeling and spectral imaging. Proc Nat Acad Sci U S A 108: 4152–4157

10. Daims H, Lücker S, Wagner M (2006) daime, a novel image analysis program for microbial ecology and biofilm research. Environ Microbiol 8:200–213

11. Schramm A, Fuchs BM, Nielsen JL, Tonnolla M, Stahl DA (2002) Fluorescence in situ hybridization of 16S rRNA gene clones (Clone-FISH) for probe validation and screening of clone libraries. Environ Microbiol 4: 713–720

Chapter 14

The Characterization of Living Bacterial Colonies Using Nanospray Desorption Electrospray Ionization Mass Spectrometry

Brandi S. Heath, Matthew J. Marshall, and Julia Laskin

Abstract

Nanospray desorption electrospray ionization (nano-DESI) coupled with high-resolution mass spectrometry (MS) and tandem mass spectrometry (MS/MS) enable detailed molecular characterization of living bacterial colonies directly from nutrient agar. The ability to detect molecular signatures of living microbial communities is important for investigating metabolic exchange between species without affecting the viability of the colonies. We describe the protocol for bacterial growth, sample preparation, ambient profiling, and data analysis of microbial communities using nano-DESI MS.

Key words Microbial communities, Nanospray desorption electrospray ionization (nano-DESI), High-resolution mass spectrometry, Metabolites

1 Introduction

Mass spectrometry (MS) enables highly specific and sensitive chemical analysis of complex molecules in biological samples. An increasingly studied topic in the MS community is the use of spatially localized sampling from surfaces for studying metabolic exchange between microbial communities grown on agar plates [1–4] Microbial biofilms produced on surfaces have been traditionally investigated using secondary ion MS (SIMS), matrix-assisted laser desorption ionization MS (MALDI-MS) [2, 5–9], or laser desorption postionization MS (LDPI-MS) [10, 11]. Given the need for sample analysis inside the vacuum system, these techniques require significant sample preparation during which the colony loses its viability [12]. Several ambient ionization techniques have been used for chemical analysis of bacterial colonies directly from agar plates. Specifically, desorption electrospray ionization (DESI) [2, 8, 13], direct analysis in real time (DART) [8], low temperature plasma (LTP) [14], laser desorption ionization

Lianhong Sun and Wenying Shou (eds.), *Engineering and Analyzing Multicellular Systems: Methods and Protocols*,
Methods in Molecular Biology, vol. 1151, DOI 10.1007/978-1-4939-0554-6_14, © Springer Science+Business Media New York 2014

(LDI) [2], and laser ablation electrospray ionization (LAESI) [8, 9] have helped overcome some of the limitations of the vacuum-based techniques [7].

We have recently developed nanospray desorption electrospray ionization (nano-DESI)—a new ambient soft ionization technique that enables sensitive detection of molecules on solid and liquid surfaces [15]. In nano-DESI experiments, the analyte is desorbed from a surface into a liquid bridge created between two fused silica capillaries. The primary capillary supplies solvent to the sample, whereas the second capillary delivers desorbed analyte molecules to a mass spectrometer inlet and ionizes them through nanospray. Nano-DESI has been used for characterization of complex organic mixtures [16–19], crude oil [20], imaging of biological tissues [7, 21, 22], and spatially resolved analysis of living microbial communities directly from agar plates [12, 23]. Here we will describe the protocol for the analysis of bacterial colonies using nano-DESI MS to study dissimilatory reduction of iron oxide nanorods by *Shewanella oneidensis* strain MR-1 [24–26], hereafter referred to as MR-1 as an example. MR-1 possesses the respiratory versatility to respire O_2 or in the absence of O_2, respire (i.e., reduce) iron^{3+} to iron^{2+} to gain energy for cellular growth. The protocol can be readily extended to other microbial systems that can be cultured on solid substrates.

2 Materials

2.1 Bacterial Growth

1. Frozen MR-1 25 % glycerol stock (*see* **Note 1**).

2. Liquid Luria Bertani (LB) Lennox broth (*see* **Note 2**).

3. Sterile Standard 100×15 mm Petri Plates (*see* **Note 3**).

4. Solid LB Lennox agar (1.5%) plates poured to a thickness of 1.5 mm (*see* **Note 2**).

5. Spectrophotometer.

6. Adjustable pipette and pipette tips.

7. AnaeroPack Rectangular Jars (Mitsubishi Gas Chemical America, Inc., New York, NY).

8. Power Goethite nanorods, 50–150 nm in diameter and 400–1,000 nm in length (Nanostructured and Amorphous Materials, Inc., Houston, TX) (*see* **Note 4**).

2.2 Sample Preparation and Ambient Profiling

1. Razor Blade.

2. Micro-Spatula.

3. Sterile Polystyrene Slides.

4. Fused Silica Capillaries (Polymicro Technologies, Phoenix, AZ).

5. U-231x Ferrule-Calco Compatible (Upchurch Scientific, Radnor, PA).

6. U-322 Union Assay ZDV Valco Type (Upchurch Scientific, Radnor, PA).

7. PEEK Polymer tubing (IDEX Health & Science LLC, Oak Harbor, WA).

8. An automated XYZ positioning stage composed of three MFA series miniature linear stages (Newport Corporation, Irvine, CA) for controlling the position of a sample holder (*see* **Note 5**).

9. XYZ500TIM micromanipulator (Quarter Research and Development, Bend, OR) for controlling the position of the primary capillary (*see* **Note 5**).

10. T12 miniature XYZ stage (Thorlabs, Newton, NJ) for controlling the position of the secondary capillary (*see* **Note 5**).

11. Thermo Finnigan LTQ/Orbitrap mass spectrometer (San Jose, CA) (*see* **Note 6**).

12. Solvent: 0.05 % formic acid, acetonitrile, toluene, and methanol (35:15:50 v/v) (*see* **Note 7**).

13. Three AD7013MTL Dino-Lite Premier Cameras (Dino-Lite Digital Microscope, Torrance, CA) (*see* **Note 8**).

2.3 Data Analysis

1. Decon2LS (PNNL, http://ncrr.pnl.gov/softwae/) (*see* **Note 9**).

2. Microsoft Excel macros and Alignment Software (developed in house) (*see* **Note 9**).

3. MIDAS Molecular Formula Calculator (http://magnet.fsu.edu/~midas/).

4. Metabolite databases: ChemSpider database (http://chemspider.com), Metlin (http://metlin.scripps.edu), MetaCyc (http://metacyc.org), Lipid MAPS (http://www.lipidmaps.org/).

3   Methods

3.1 Bacterial Growth

1. Suspend an aliquot of 10 µL of frozen bacterial stock into 3–5 mL of liquid culture.

2. Grow aerobically at 30 °C while shaking at 100 rpm to an optical density (OD_{600}) of 0.3, as measured with a spectrophotometer.

3. Suspend nanorods (used as a reducing source for cells) in sterile water and store in a glass vial covered with aluminum foil.

4. Vortex nanorods.

5. Inoculate 20 µL of liquid culture onto agar plates in lines (*see* **Note 10**). The bacterial colonies can be inoculated to grow in a variety of geometries (*see* Table 1).

6. Pipette an aliquot of 20 µL of the suspended nanorods in a line 1.5 cm away from the bacterial colony. This distance was

Table 1
Selected inoculation geometries for the analysis of a bacterial colony (region highlighted by the blue dashed line) relative to another species (region highlighted by the red dashed line). Selected traces of sampling by nano-DESI are shown as yellow dashed lines. There are many options for the inoculation of the bacterial colonies; several are shown here. Secondary species can be a nanorods iron source or another bacterium

Geometry	Picture	Description
Parallel lines I	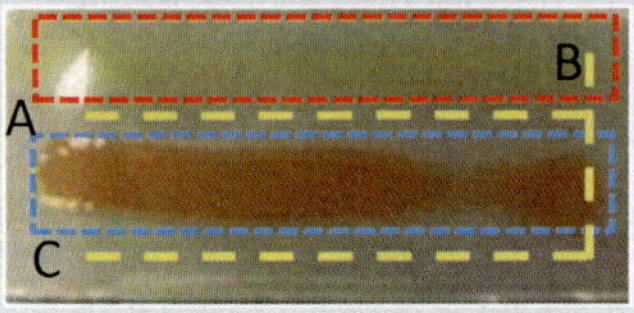 A B C	Inoculation of bacterial colonies in parallel lines allows for a large area to sample from with the nano-DESI probe. Useful for long data acquisitions such as automated MSMS experiments. Trace A enables sampling of metabolites between the two species. Trace B provides a cross-sectional view of the interacting system. Trace C enables detection of metabolites produced by the colony without any interaction, providing a chemical signature of the control sample on the same agar plate
Parallel lines II	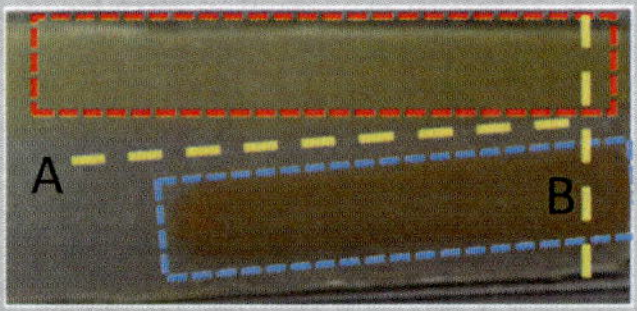 A B	The extended line of one species can be used as a control and allows for additional characterization of the sample prior to nano-DESI analysis. For example, in our studies the extension enabled the detection of metal reduction activity close and far away from the colony
Line and dot	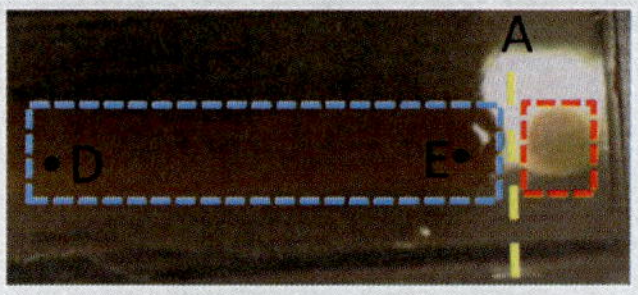 A D E	The two species can be placed in a line (blue) and dot (red) formation. Spectra collected near dots D and E represent bacterial signatures from the colony itself (control) and the side of the colony close to the other species. Trace A samples the interaction region
Diagonal line	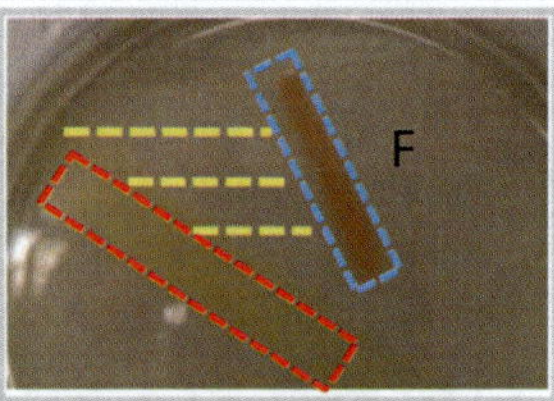 F	This geometry provides useful information on the dependence of chemical gradients on the distance between the interacting species. Sampling along traces F generates a two-dimensional representation of the chemical gradients
Zig–zag	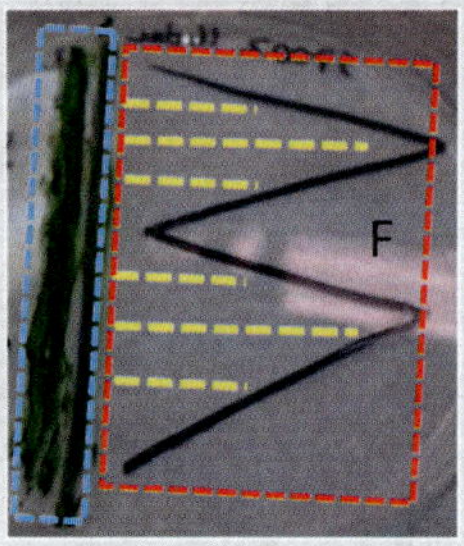 F	This is an extension of the diagonal line geometry, in which the second colony is grown in a zig–zag configuration (solid black line). This geometry enables multiple sampling of chemical gradients (traces F) as a function of the distance between the two species on the same agar plate

chosen to ensure better control of the sample preparation while keeping the distance between the iron oxide source and the colony as small as possible.

7. Seal each plate with parafilm to prevent evaporation.

8. Grow the samples aerobically at 30 °C for 24 h to allow for biomass accumulation (*see* **Note 11**).

9. Place the samples in anaerobic jars to observe metabolism in an anaerobic environment. Samples were transferred into an anaerobic glove bag containing a nitrogen/hydrogen (95 %/5 %) or equivalent anoxic atmosphere (seal anaerobic jars in this atmosphere and incubate samples at room temperature (~21 °C) until they are ready to be sampled).

10. Grow the samples in anaerobic conditions for a period of 3 days (*see* **Note 12**).

3.2 Ambient Profiling Using Nano-DESI MS

1. Make the solvent of choice for analysis. In this example we used a mixture of 0.05 % formic acid, acetonitrile, toluene, and methanol (35:15:50 v/v) as a working nano-DESI solvent.

2. Make a nano-DESI probe out of fused silica capillaries. Figure 1 shows a schematic drawing of the custom-designed nano-DESI source. The probe is composed of two capillaries—primary and nanospray. The primary capillary delivers solvent to the nanospray capillary forming a liquid bridge. When high

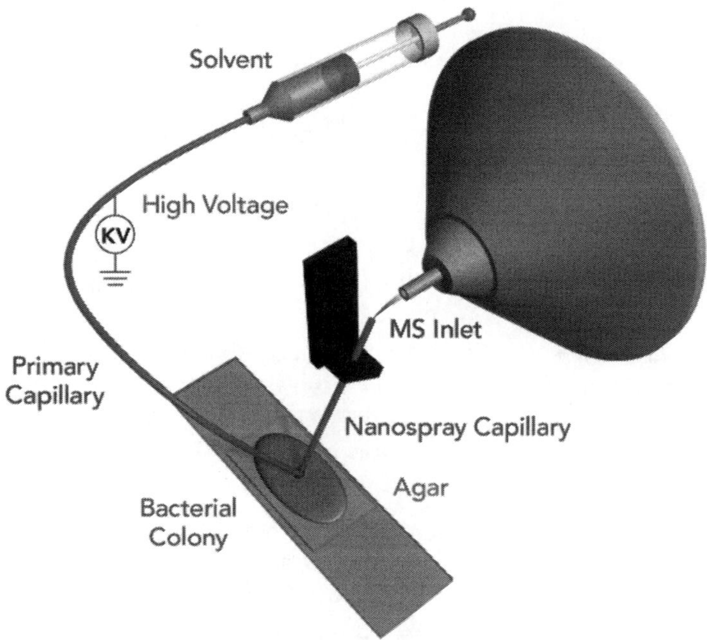

Fig. 1 Schematic drawing of the nano-DESI setup for sampling and spatial profiling of microbial communities directly from agar plates

voltage is applied, the nanospray capillary empties into the mass spectrometer inlet by way of electrospray.

3. Position capillaries using the XYZ micromanipulators so that the primary and secondary capillaries are touching. Observe the relative position of the capillaries using Dino-Lite cameras (*see* **Note 8**).

4. Connect the primary capillary to a syringe pump and propel liquid through the primary capillary forming a liquid bridge between the primary capillary and the nanospray capillary (*see* Fig. 1).

5. Set your syringe pump to a constant volumetric flow to create a liquid bridge between the two capillaries (*see* **Note 13**). Adjust the position of the nanospray capillary relative to the mass spectrometer inlet to maintain the liquid bridge as needed.

6. Apply a high voltage and adjust the flow rate and the position of the nanospray capillary relative to the inlet to achieve a stable electrospray from the nanospray capillary into the heated capillary (*see* **Note 14**).

7. Cut the area of interest from the petri dish using a sterile razor blade and micro-spatula. Place them on sterile polystyrene slides for nano-DESI analysis (*see* **Note 15**).

8. Mount the slide on the automated XYZ stage.

9. Raise the sample until it is in contact with the nano-DESI probe and adjust the distance between the probe and the sample to obtain stable ion signal.

10. Collect mass spectra by placing the probe onto selected spots on the sample or by running the sample in preset line scans under the nano-DESI probe (*see* **Note 16**). Table 1 shows selected traces of sampling options for nano-DESI probe.

11. Collect tandem mass spec data (MS/MS) by isolating the precursor ion of interest.

12. Ramp the collision energy manually from 0 to 25 until the precursor ion is completely gone.

13. Repeat for all peaks of importance, saving each file.

3.3 Data Analysis
(See Note 17)

1. Extract mass spectral features with a desired signal-to-noise ratio from the mass spectra using Decon2LS software. In this study, we obtained average mass spectra over the entire acquisition and then used signal-to-noise of five for peak picking (*see* **Note 9**).

2. Align mass spectra obtained from different samples or different locations on the same sample within a specified tolerance (*see* **Note 9**).

3. Remove the peaks corresponding to the ^{13}C isotopes from the list by examining m/z values separated by the mass

difference between the masses of ^{13}C and ^{12}C (1.00335). Note that the relative abundance of the peak corresponding to the ^{13}C should be consistent with the calculated isotopic distribution at a specific m/z value.

4. Eliminate solvent, agar and noise peaks from the data set. These peaks can be obtained from the agar-only control.

5. Identify peaks as described in detail elsewhere [27] (*see* **Note 18**). Briefly, perform a second-order mass defect transformation on the list of m/z values using CH_2 and H_2 as bases. Cluster the two-dimensional homologous series of peaks separated by the number of CH_2 and H_2 units into distinctive groups. Identify one member in each group. This will allow you to identify each member of the group.

6. Enter exact masses of one member of each group into the MIDAS Molecular Formula Calculator (http://www.magnet.fsu.edu/~midas/) to assign elemental compositions of the molecules.

7. Screen the elemental compositions against the ChemSpider (http://chemspider.com) [28], Metlin (http://metlin.scripps.edu) [29], LIPID MAPS [30], and MetaCyc (http://metacyc.org) [31] databases.

8. Create MS/MS spectra that contain both the precursor ion and the fragment ions.

9. Compare MS/MS spectra with data found in databases and literature to confirm identities of molecules.

---

4   Notes

1. Use the frozen stock of the bacteria of interest.

2. Use the optimal broth and agar for the bacteria of interest.

3. Smaller or larger petri plates can also be used.

4. For studying interactions between two or more different bacterial colonies, repeat **steps 1–3** for each of the other bacteria. For studying interaction between a colony and an inanimate object such as nanorod iron source, deposit that object on the agar plate.

5. Any positioning devices can be used.

6. Any mass spectrometer, high or low resolution, can be used.

7. The solvent can be optimized to detect molecules of interest.

8. Any camera can be used. The Dino-Lite camera has approximately 90× magnification at a working distance of ~5 cm.

9. Any peak picking and alignment programs can be employed here.

10. For investigating two or more colonies or the interaction between a colony and an inanimate object inoculate cells and/or place object on the agar plate during this step.

11. If you are growing a strictly anaerobe bacteria, skip this step. Grow your samples at the optimum temperature for that particular bacterium.

12. Bacteria grow at different rates, temperatures, and in the presence of other stimuli, such as light or oxygen concentration or media composition. Grow your samples for the specific time and in the specific experimental conditions that will elicit the desired phenotype for that bacterium.

13. Typical volumetric flow is between 0.5 and 2.5 μL/s. This value will increase or decrease based on the position of the nanospray capillary relative to the heated capillary. Typical capillary size is 150 μm OD $\times$ 50 μm ID.

14. Typical high voltage was between 2.0 and 3.0 kV in this study.

15. This step is only necessary if the sides of the petri dish interfere with the mass spectrometer inlet. If analysis can be performed without removing the samples, skip this step.

16. The analysis can be performed in positive and/or negative ionization mode.

17. For data analysis of spatially resolved experiments refer to ref. [32].

18. There are many other approaches for processing the data.

Acknowledgements

The research described in this paper is part of the Chemical Imaging Initiative (CII) at Pacific Northwest National Laboratory (PNNL). It was conducted under the Laboratory Directed Research and Development (LDRD) Program at PNNL, a multiprogram national laboratory operated by Battelle for the U.S. Department of Energy (DOE). J.L., B.S.H., M.J.M. acknowledge support from the CII at PNNL. B.S.H. was supported in part by the DOE Science Undergraduate Laboratory Internship (SULI) at PNNL. The authors acknowledge fruitful discussions with Drs. Jim Fredrickson and Margie Romine (PNNL). A portion of this research was performed using EMSL, a national scientific user facility sponsored by DOE Office of Biological and Environmental Research (OBER) and located at PNNL. PNNL is operated by Battelle for DOE.

References

1. Esquenazi E, Yang Y-L, Watrous J, Gerwick WH, Dorrestein PC (2009) Imaging mass spectrometry of natural products. Nat Prod Rep 26:1521

2. Watrous JD, Dorrestein PC (2011) Imaging mass spectrometry in microbiology. Nat Rev Microbiol 9:683–694

3. Moree WJ, Phelan VV, Wu CH, Bandeira N, Cornett DS, Duggan BM et al (2012) Interkingdom metabolic transformations captured by microbial imaging mass spectrometry. Proc Natl Acad Sci 109:13811–13816

4. Phelan VV, Liu W-T, Pogliano K, Dorrestein PC (2012) Microbial metabolic exchange–the chemotype-to-phenotype link. Nat Chem Biol 8:26–35

5. Cornett DS, Reyzer ML, Chaurand P, Caprioli RM (2007) MALDI imaging mass spectrometry: molecular snapshots of biochemical systems. Nat Methods 4:828–833

6. Van Baar BLM (2000) Characterisation of bacteria by matrix-assisted laser desorption/ionisation and electrospray mass spectrometry. FEMS Microbiol Rev 24:193–219

7. Laskin J, Heath BS, Roach PJ, Cazares L, Semmes OJ (2012) Tissue imaging using nanospray desorption electrospray ionization mass spectrometry. Anal Chem 84:141–148

8. Vertes A, Hitchins V, Phillips KS (2012) Analytical challenges of microbial biofilms on medical devices. Anal Chem 84:3858–3866

9. Parsiegla G, Shrestha B, Carrière F, Vertes A (2012) Direct analysis of phycobilisomal antenna proteins and metabolites in small cyanobacterial populations by laser ablation electrospray ionization mass spectrometry. Anal Chem 84:34–38

10. Akhmetov A, Moore JF, Gasper GL, Koin PJ, Hanley L (2010) Laser desorption postionization for imaging MS of biological material. J Mass Spectrom 45:137–145

11. Gasper GL, Carlson R, Akhmetov A, Moore JF, Hanley L (2008) Laser desorption 7.87 eV postionization mass spectrometry of antibiotics in *Staphylococcus epidermidis* bacterial biofilms. Proteomics 8:3816–3821

12. Watrous J, Roach P, Alexandrov T, Heath BS, Yang JY, Kersten RD et al (2012) Mass spectral molecular networking of living microbial colonies. Proc Natl Acad Sci U S A 109(26): E1743–E1752

13. Zhang JI, Talaty N, Costa AB, Xia Y, Tao WA, Bell R et al (2011) Rapid direct lipid profiling of bacteria using desorption electrospray ionization mass spectrometry. Int J Mass Spectrom 301:37–44

14. Zhang JI, Costa AB, Tao WA, Cooks RG (2011) Direct detection of fatty acid ethyl esters using low temperature plasma (LTP) ambient ionization mass spectrometry for rapid bacterial differentiation. Analyst 136:3091

15. Roach PJ, Laskin J, Laskin A (2010) Nanospray desorption electrospray ionization: an ambient method for liquid-extraction surface sampling in mass spectrometry. Analyst 135:2233–2236

16. Laskin J, Eckert PA, Roach PJ, Heath BS, Nizkorodov SA, Laskin A (2012) Chemical analysis of complex organic mixtures using reactive nanospray desorption electrospray ionization mass spectrometry. Anal Chem 84: 7179–7187

17. Roach PJ, Laskin J, Laskin A (2010) Molecular characterization of organic aerosols using nanospray-desorption/electrospray ionization-mass spectrometry. Anal Chem 82:7979–7986

18. Liu P, Lanekoff IT, Laskin J, Dewald HD, Chen H (2012) Study of electrochemical reactions using nanospray desorption electrospray ionization mass spectrometry. Anal Chem 84:5737–5743

19. Bateman AP, Laskin J, Laskin A, Nizkorodov SA (2012) Applications of high-resolution electrospray ionization mass spectrometry to measurements of average oxygen to carbon ratios in secondary organic aerosols. Environ Sci Technol 46:8315–8324

20. Eckert PA, Roach PJ, Laskin A, Laskin J (2012) Chemical characterization of crude petroleum using nanospray desorption electrospray ionization coupled with high-resolution mass spectrometry. Anal Chem 84:1517–1525

21. Lanekoff I, Heath BS, Liyu A, Thomas M, Carson JP, Laskin J (2012) Automated platform for high-resolution tissue imaging using nanospray desorption electrospray ionization mass spectrometry. Anal Chem 84:8351–8356

22. Lanekoff I, Thomas M, Carson JP, Smith JN, Timchalk C, Laskin J (2013) Imaging nicotine in rat brain tissue by use of nanospray desorption electrospray ionization mass spectrometry. Anal Chem 85:882–889

23. Lanekoff I, Geydebrekht O, Pinchuk GE, Konopka AE, Laskin J (2013) Spatially resolved analysis of glycolipids and metabolites in living Synechococcus sp. PCC 7002 using nanospray desorption electrospray ionization. Analyst 138(7):1971–1978

24. Fredrickson JK, Romine MF, Beliaev AS, Auchtung JM, Driscoll ME, Gardner TS et al (2008) Towards environmental systems biology of *Shewanella*. Nat Rev Microbiol 6:592–603

25. Tang YJ, Meadows AL, Kirby J, Keasling JD (2006) Anaerobic central metabolic pathways in *Shewanella oneidensis* MR-1 reinterpreted in the light of isotopic metabolite labeling. J Bacteriol 189:894–901

26. Maier TM, Myers CR (2004) The outer membrane protein Omp35 affects the reduction of Fe(III), nitrate, and fumarate by *Shewanella oneidensis* MR-1. BMC Microbiol 4:23

27. Roach PJ, Laskin J, Laskin A (2011) Higher-order mass defect analysis for mass spectra of complex organic mixtures. Anal Chem 83:4924–4929

28. Little JL, Williams AJ, Pshenichnov A, Tkachenko V (2011) Identification of "known unknowns" utilizing accurate mass data and ChemSpider. J Am Soc Mass Spectrom 23:179–185

29. Smith CA, O'maille G, Want EJ, Chuan Q, Trauger SA, Brandon TR et al (2005) METLIN: a metabolite mass spectral database. Ther Drug Monit 27:747–751

30. Cotter D (2006) LMPD: LIPID MAPS proteome database. Nucleic Acids Res 34:D507–D510

31. Caspi R, Altman T, Dale JM, Dreher K, Fulcher CA, Gilham F et al (2009) The MetaCyc database of metabolic pathways and enzymes and the BioCyc collection of pathway/genome databases. Nucleic Acids Res 38:D473–D479

32. Thomas M, Heath BS, Laskin J, Li D, Liu E, Hui K et al (2012) Visualization of high resolution spatial mass spectrometric data during acquisition. Conf Proc IEEE Eng Med Biol Soc 2012:5545–5548

Chapter 15

Modeling Community Population Dynamics with the Open-Source Language R

Robin Green and Wenying Shou

Abstract

The ability to explain biological phenomena with mathematics and to generate predictions from mathematical models is critical for understanding and controlling natural systems. Concurrently, the rise in open-source software has greatly increased the ease at which researchers can implement their own mathematical models. With a reasonably sound understanding of mathematics and programming skills, a researcher can quickly and easily use such tools for their own work. The purpose of this chapter is to expose the reader to one such tool, the open-source programming language R, and to demonstrate its practical application to studying population dynamics. We use the Lotka–Volterra predator–prey dynamics as an example.

Key words Modeling, R, Lotka–Volterra, Population dynamics, Predator–prey relationship

1   Introduction

Mathematics is integral to the study of biological systems. From the direct application of the Malthusian growth model [1] to abstraction from Fibonacci number series, mathematical models can help researchers explain natural phenomena quantitatively and generate new hypotheses better than with only experimental observations. After translating a biological problem into a set of mathematical equations, solutions can be sought and visualized.

Perhaps one of the most popular tools for such analysis is the open-source language and computing environment R (r-project.org). First developed by Ross Ihaka and Robert Gentleman at the University of Auckland in 1993, R is part of the Free Software Foundation's GNU Project, a massive collaborative effort meant to develop high quality open-source software (http://www.fsf.org/). R offers users a plethora of standard statistical and computational tools, extensive collections of predefined functions, and a well-maintained and documented support system. In addition to the preexisting functionalities, R also allows users to define their own functions

Lianhong Sun and Wenying Shou (eds.), *Engineering and Analyzing Multicellular Systems: Methods and Protocols,*
Methods in Molecular Biology, vol. 1151, DOI 10.1007/978-1-4939-0554-6_15, © Springer Science+Business Media New York 2014

and algorithms. Syntax (jargon for rules and structure of the programming language) of R is also relatively easy to understand.

Here we will demonstrate how R can be used to express and analyze mathematical models of population dynamics. We describe the Lotka–Volterra equations for representing population dynamics between predator and prey. We then present a step-by-step guide to getting set up to use the R environment, and an easy-to-follow implementation of the above model in R. By the end of this chapter, the reader will have a basic understanding of how to implement and numerically solve a mathematical model based on differential equations, visualize the solutions, and explore different permutations to formulate new hypotheses.

> Disclaimer: The reader should note that this chapter is not intended to give a full background or tutorial on R. For a more comprehensive introduction to R, please *see* ref. 2. It should also be noted that R, at its current stage, may have a slower performance than other languages for specific types of problems.

2   Background: The Lotka–Volterra Equations

A fundamental phenomenon in population ecology is predation, the feeding of one organism (the predator) on another (the prey). In 1926, the biophysicist Alfred Lotka proposed a mathematical model [3] to represent this relationship. The Italian mathematician Vito Volterra explored this relationship independently of Lotka [4]. This has led to the proposal of the *Lotka–Volterra equations*:

$$\frac{dx}{dt} = Ax - Bxy. \tag{1}$$

$$\frac{dy}{dt} = Cxy - Dy. \tag{2}$$

where

- $x(t)$ is population density of the prey at time t.
- $y(t)$ is the population density of the predator at time t.
- A is net birth rate (natural birth rate subtracting natural death rate, in the unit of per time unit) of the prey population in the absence of predator.
- B is the rate at which prey are killed due to the presence of predator (in the unit of per time unit per predator density).
- C is the birth rate of the predator population due to the presence of prey (in the unit of per time unit per prey density).
- D is the death rate of the predator population in the absence of prey.

- dx/dt and dy/dt are the rates of change of x and y, respectively.

Intuitively, the reader can think of the two differential equations as:

The rate at which the prey population changes is the birth rate of the prey minus the rate of consumption of the prey by the predator

and

The rate at which the predator population changes is the birth rate of the predator (which is dependent on the amount of prey present) minus the death rate of the predator.

Thus these rates are dependent on the densities of both the predator and the prey populations, in addition to parameters which are static in this model.

It is important to note that this model does make assumptions that might not necessarily be true:

- There is an ample source of food for the prey at all times.

- The predator population only feeds on the prey population (no other source of food) and feeds continuously.

- There is infinite space to hold both predator and prey populations.

- The rate of change of the population is proportional to its density.

- The interactions between predator and prey are determined by the product of the density of the two populations, much like in the collision of two reactants in concerted bimolecular reactions. There are no spatial refuges for prey.

From Eqs. 1 and 2 at time $t' = t + dt$, where dt approaches 0,

$$x(t') = x(t) + \left(\frac{dx}{dt} \times dt\right) = x(t) + (Ax - Bxy) \times dt. \qquad (3)$$

$$y(t') = y(t) + \left(\frac{dy}{dt} \times dt\right) = y(t) + (Cxy - Dy) \times dt. \qquad (4)$$

Once the initial $(t = 0)$ values for $x(t)$ and $y(t)$ are known, the prey and predator populations densities $x(t)$ and $y(t)$ can be computed for any t.

For pedagogical purposes, we first present an R implementation based on Euler (first-order) approximations in Eqs. 3 and 4 to estimate population dynamics corresponding to Eqs. 1 and 2. We will then present a more practical and efficient implementation using the *deSolver* package to solve Eqs. 1 and 2. For a more complete overview, please *see* ref. 5.

3   Getting Started with R

For this chapter, the authors ran all simulations in RStudio (www. rstudio.com), an open-source environment for R computing. While certainly not the only environment available, RStudio is simple and provides an integrated environment for basic computation, writing scripts, and visualizing data in addition to up-to-date documentation on various aspects of the language.

To install RStudio, please visit www.rstudio.com/ide/download/desktop and download the package most suitable for your operating system (the authors recommend you select the version under "Recommended For Your System" at the top of the page). Once the package is downloaded, click on the file and follow the on-screen instructions to install all files in the proper directories. Depending on where you chose to install RStudio, the graphical user interface (GUI) icon should appear in that directory. Click on it and you should see a screen similar to the one below (*see* Fig. 1):

The first thing to do is to familiarize yourself with the environment. The "Console" window is where you can perform simple computations, call scripts and functions, and create variables for later use. For example, in the "Console" window, type the following commands ("$>$" automatically appears in the Console for a new command) (*see* Fig. 1):

```
>a<-2
>b<-50
```

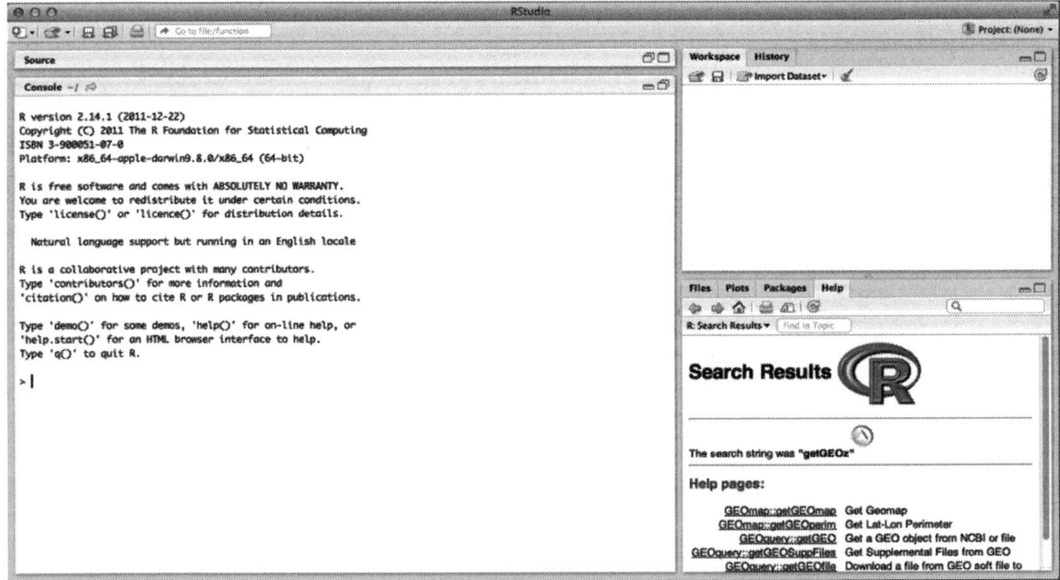

Fig. 1 New RStudio session

This is called "defining a variable." Basically, you are telling the current environment that the character *a* now holds the value (in R, this value is called a "numeric") 2, and the character *b* holds the value '50. Next, type the following command (note the quotation marks):

```
>word<- 'airplane'
```

The assignment of variable names is not limited to single letters or symbols (something we will exploit later in our code). In this case, you are telling the environment that *word* holds the string of characters (in R, this is known simply as a "character") that make up "airplane".

These variables that you have created can be manipulated. Type the following commands:

```
> a+b
[1] 52
> a-b
[1] -48
> a*b
[1] 100
> b/a
[1] 25
>
```

Your screen should show the values 52, –48, 100, and 25. As you can see, simple arithmetic operations can be performed with the newly created variables.

Next, the user should become familiar with their working directory. In your console, type the following command to get the current working directory:

```
> getwd()
```

You should see something equivalent to the following:

```
> getwd()
[1] "/Users/user1"
```

This location can be changed with the following command:

```
> setwd('path/to/desired/directory')
```

To save your work to the current working directory, type the following command:

```
> save.image()
```

This will create a file called ".RData" (this will be a hidden file in most directories). To load the ".RData" file for future use, type the following command:

```
> load('.RData')
```

Alternatively, RData can be files that are explicitly named during the save process, which also makes them visible in directories:

```
>save.image('LotkaVolterraExample.RData')
>load('LotkaVolterraExample.RData')
```

Use the up arrow key to return to the previous command, and repeat this process to access earlier commands. Now that you are familiar with the basics, you are ready to begin implementing the Lotka–Volterra equations.

4   Implementation

To get started, click on the "File" button in the top left-hand corner of the screen and select "File->New->R-Script". This should be a drop down screen in RStudio that looks something like the following (Fig. 2):

In this new window, type the following (Fig. 3):

This is the framework that will contain the *function* that you will write. Inputs to a function can be put inside ( ). Next, set up your environment to easily save and run your work. To "tell" the R environment to use the code you have written, you may *"source"* your script by selecting "Source on Save" from the top left corner of your window, as shown below (Fig. 4):

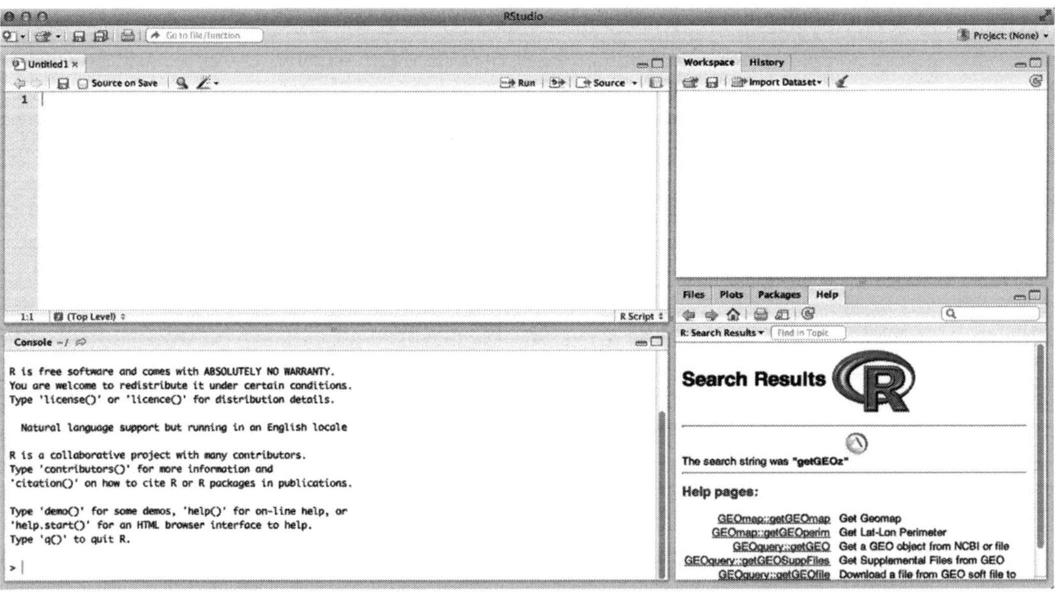

Fig. 2 New RScript window

Fig. 3 New Lotka–Volterra function

Fig. 4 Source on save command

Sourcing can be thought of as a way of making the R environment reevaluate the file/script in question. By sourcing the file/script, you are telling the R environment to execute the file/script, which can either result in running a program or in this case, updating a function. Next, click on the floppy disc icon to save your script as, for example, LV_Example. Since you are saving your script for the first time, name the file as you wish (it will be saved as a .R file). This file will be saved in your current working directory. When you save your file, on your Console screen you should see something similar to the following output:

```
> source('~/LV_Example.R')
```

This means your function is now "ready" to be called in your environment. Add the following print statement to your function in the scripting window and save/source your script (Fig. 5):

Now "call" your function by typing the following:

```
> Lotka_Volterra()
```

You should see the following on your screen:

```
[1] "This statement will be printed"
```

Note that we added the print command in between the two curly braces ("{ }") of the function. Anything written in between these curly braces will only be executed when the function is called. If the same print command was instead placed outside the braces, then the command would be executed every time the file is sourced.

```
Lotka_Volterra.R ×
Source on Save
1  Lotka_Volterra<-function()
2  {
3
4    print('This statement will be printed')
5
6
7
8
9  }
```

Fig. 5 Adding a print command

Next, we will describe how to pass variables to your function. Add the following commands to your function:

```
Lotka_Volterra<-function(x_start)
{
   print('This statement will be printed')
   print(x_start)
}
```

Now, call your function like before. You will notice an error message:

```
> Lotka_Volterra ()
[1] "This statement will be printed"
Error in print(x_start): argument "x_start" is
missing, with no default
```

Because no value was passed to the function, yet the function required a value to be assigned to "x_start", we got an error message. This can be solved in two ways. The first is to explicitly pass a value to the function:

```
> Lotka_Volterra(8)
[1] "This statement will be printed
[1] 8
> Lotka_Volterra(x_start=8)
[1] "This statement will be printed"
[1] 8
```

Or, alternatively, you can define a *default value* for "x_start" in your RScript function, shown below:

```
Lotka_Volterra<-function(x_start=10)
{
```

This means that if no value is passed to the function, "x_start" will automatically be assigned to 10. However, explicitly passing a value to the function overrides this:

```
> Lotka_Volterra()
[1] "This statement will be printed"
[1] 10
> Lotka_Volterra(8)
```

```
[1] "This statement will be printed"
[1] 8
> Lotka_Volterra(x_start=8)
[1] "This statement will be printed"
[1] 8
```

It may be easier to define default parameters for the purpose of this program; otherwise calling the function each time will require you to specify every parameter in the console, which can be onerous.

Modify your function as follows (be sure to delete the two previous print statements you were using within your function, as they are no longer needed):

```
Lotka_Volterra<-function(x_start=5,y_start=2,A=
1,B=0.2,C=0.04,D=0.5,iterations=10000,t
ime_step=0.01)
```

These will be the default parameters for running the simulations. It may be useful to keep a record of what each of these parameters represents when writing the code. This can be accomplished by *commenting* our code. Comments are not "read" by the program when running, so they only serve as documentation for the programmer (or someone else reading the code in the future). Commenting in R can be accomplished by adding "#" at the beginning of a line in our script. We can use comments to add notes about our code, as shown below (Fig. 6):

Next, add the following commands to your code:

```
x<-x_start
y<-y_start

dt<-time_step
graph_frame<-c(0,x,y)
```

```
Lotka_Volterra<-function(x_start=5,y_start=2,A=1,B=0.2,C=.04,D=0.5,iterations=10000,time_step=0.01)
{

   # x_start= Starting Density of Prey
   # y_start= Starting Density of Predators
   # x= Density of Prey
   # y= Density of Predators
   # x_new = Reestimation of x for each iteration
   # y_new = Reestimation of y for each iteration
   # A= Net Birth Rate of Prey Population
   # B= Rate of Consumption of Prey by Predator
   # C= Birth Rate of the Predator Population due to Consumption of Prey
   # D= Death rate of the Predator Population in the Absence of Prey
   # iterations= Number of estimations to make
   # time_step= time step (dt) needed for Euler's approximation
```

Fig. 6 Documenting the code

The first two commands simply assign the values of "x_start" and "y_start" to the new variables "x" and "y", respectively. The third command sets the time step for simulation "dt" to the value stored in "time_step." The last command creates a *vector* called "graph_frame." The vector is one of the most important data types in R. Intuitively, it can be thought of as a collection of variables held together in a searchable container (similar to lists in Python or arrays in C++). Essentially, we have created a 1×3 table of values, which will be expanded to store the results of our code. The first column will hold information on time (which starts at 0 and is in an arbitrary unit), the second and third columns a will respectively hold the population densities of prey and predator corresponding to the time point in the first column.

Next, add the following to your code, just below the above commands:

```
for( i in 1:iterations)
{



}
```

This is known as defining a *loop*. A loop is a section of code that is repeated until some criterion tells the loop to stop or a command is explicitly given to break out of the loop. In our case, we have qualified that our loop should continue until the "for" statement is no longer true. In our "for" statement, we are implicitly defining a vector containing all the numbers from one to *iterations*. Intuitively, this can be thought of as saying:

> Repeat the loop until the value of i, which iterates through a vector of all the number of iterations, is equal to the number specified in the variable *iterations* (in this case, has reached the end of the vector).

It is important to note that the value of i will increase by one as it traverses the vector of values from one to the number of iterations. To test this, add the following print command to your loop:

```
for( i in 1:iterations)
{
    print(i)

}
```

If you source your code and call the function, you'll notice that, new lines of numbers ranging from 1 to 10,000 appear on your console, demonstrating that "i" is being increasing while the code stays within the loop. Now you have the entire framework in place to begin modeling with the Lotka–Volterra equations.

Start by adding the code required to update the population densities of predator and prey at each iteration:

```
for( i in 1:iterations)
{

    dx<-(A*x-B*x*y)*dt
    dy<-(C*y*x-D*y)*dt

    x_new<-x+dx
    y_new<-y+dy

    print(i)

}
```

Within each iteration, dx and dy calculate changes in population densities of x and y according to Eqs. 3 and 4, respectively. New values of x and y after time dt are evaluated and stored in variables "*x_new*" and "*y_new*", respectively.

We then add the following lines in the loop after the definition of *x_new* and *y_new* inside the loop:

```
new_data<-c(i*dt,x_new,y_new)

graph_frame<-rbind(graph_frame,new_data)
```

The first line stores three pieces of information (time elapsed $i \times$ dt, and the values of "*x_new*" and "*y_new*" at time $i \times$ dt) in a vector called *new_data*. The next line adds this *new_data* to our *graph_frame* as a new row through a command called *rbind* (i.e. row-bind), expanding an original $N \times 3$ table to an $(N+1) \times 3$ data vector (which can also be referred to as a *matrix*).

Finally, update the values of "*x*" and "*y*" for the next iteration:

```
x<-x_new
y<-y_new
```

Delete the "print(i)" at the end of the code and your loop should loop like the one below (Fig. 7):

Thinking about the code within the loop logically, we can summarize the sequence of events as follows:

For each iteration/given unit of time:

- estimate changes in the values of x and y from the last iteration based on the Lotka-Volterra equation
- apply these changes to x and y to get new estimates
- add these new estimates and their corresponding time stamp to our overall vector
- update the values of x and y with new estimates

The result of this loop is an $N \times 3$ table/matrix of points, with each of the N rows containing a time point, and an estimate of the numbers of prey and predator for that time point.

```
for(i in 1:iterations)
{

  dx<-(A*x-B*x*y)*dt
  dy<-(C*y*x-D*y)*dt

  x_new<- x+dx
  y_new<-y+dy

  new_data<-c(i*dt,x,y)

  graph_frame<-rbind(graph_frame,new_data)

  x<-x_new
  y<-y_new

}
```

Fig. 7 Final loop code

These data can also be visualized easily in RStudio by using the "matplot" function to plot columns in a given vector against one another. Add the following to your code outside of the loop after all iterations:

```
matplot(x=graph_frame[,1],y=graph_frame
[,c(2,3)],pch=20)
```

Notice how we have selected which columns to use in the matplot function. Multidimensional vectors are indexed row by column. To select everything in the first column (for the x-axis), we leave the first entry in brackets empty and select "1" for the second entry. To select both the second and third columns to be plotted against the first column, we use a vector containing the columns we want, in this case "2" and "3". The "pch" command is simply specifying how the data should be represented, with different values specifying different shapes. Here, *pch* = 20 represents solid circles.

Now call your function. Something similar to the below should appear in the bottom right of your screen (Fig. 8):

This plot can be made more readable with modifications to the matplot function. For example, each axis can be labeled and a title can be added as follows:

```
matplot(x=graph_frame[,1],y=graph_frame[,c(2,3)
    ],pch=20,xlab='Time',ylab='Population
  Density',main='Lotka-Volterra Simulation')
```

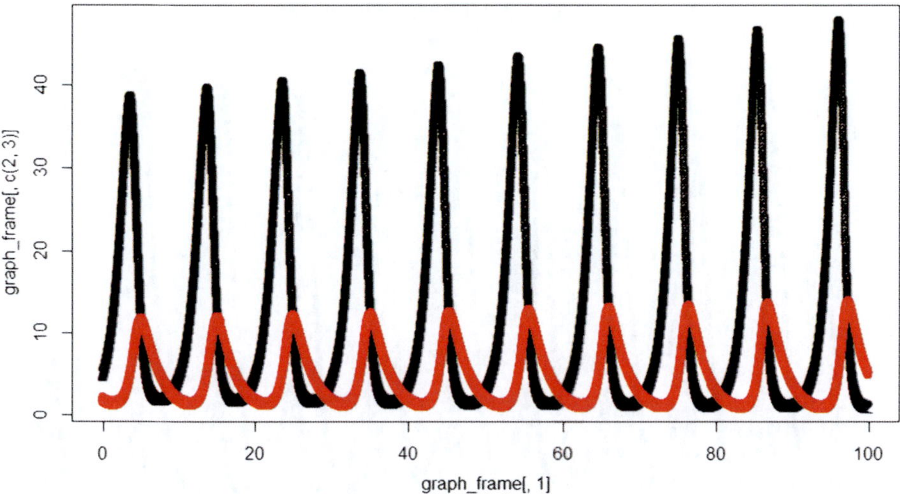

Fig. 8 Initial simulation result

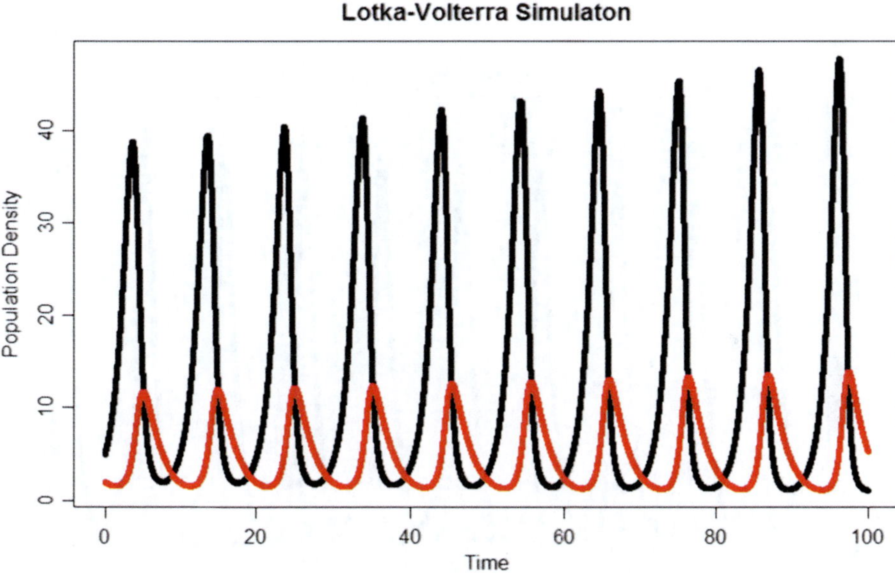

Fig. 9 Initial simulation result

This should give you a plot like this (*see* Fig. 9):

The colors of the plots can be changed as well by specifying with the "col" option in matplot (*see* Fig. 10):

```
matplot(x=graph_frame[,1],y=graph_frame[,c(2,3)],
        pch=20,xlab='Time',ylab='Population
Density',main='Lotka-Volterra Simulation',
        col=c('darkgreen','darkblue'))
```

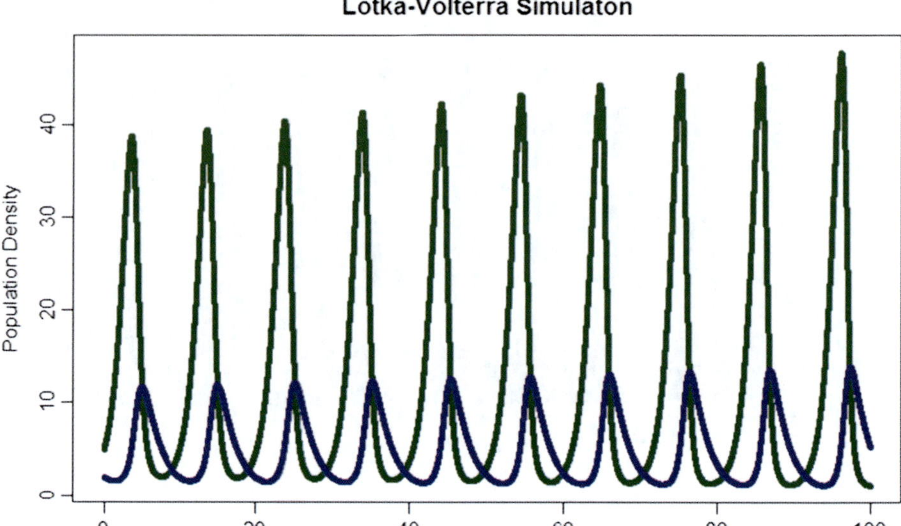

Fig. 10 Results with new colors

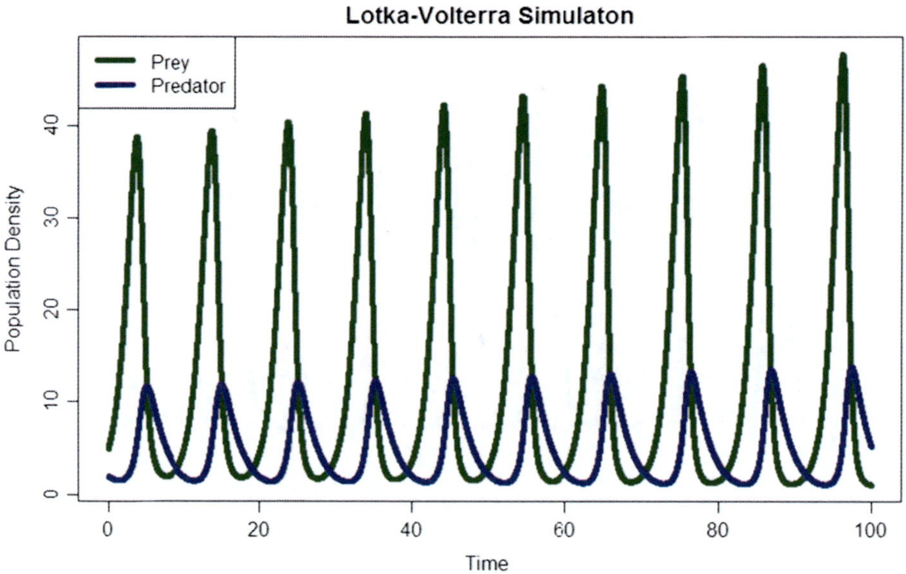

Fig. 11 Result with legend

Finally, a legend can be added to the plot for convenience (note that this is separate from the matplot function) (*see* Fig. 11):

```
legend(x='topleft',legend=c('Prey','Predator'),
       col=c('darkgreen','darkblue'),lwd=5)
```

At this point, you can delete the any early print commands, as they are not needed for further work with the program. Note: For certain calculations, loops may not be the most time-efficient method. Unless storage for the values generated in a loop are pre-allocated, calculations typically take less time using slightly more complex built-in functions in R like the apply function, which applies a given calculation over a table of values (also known as matrix calculation). For more information see the R help (type '?") for 'for', 'vector', 'matrix' and 'apply'.

5   Analysis of the Lotka–Volterra Model

Perhaps one of the most striking characteristics of the plot is its oscillatory nature. At periodic intervals, the prey and predator populations will dramatically rise and fall but out-of-phase with each other. Notice the rapid increase in the prey population density just prior to a rapid rise in the predator population density, followed by a sudden drop in the prey population density which in turn causes a drop in the predator population density. Biologically, this pattern can be explained:

For some given period of time:

- a low predator population density (which is the result of a low prey population density i.e. lack of food supply) allows the prey population to expand through reproduction with little predation
- the sudden increase in prey population density now results in an increased food supply for the predators, which allows the predators to consume and reproduce more, resulting in a larger predator population density and lower prey population density
- when the prey population density decreases, the predators no longer have a food supply to sustain their population, so they begin to die and decrease in population density

This cycle repeats infinitely in this model.

An important trend to note is the increasing peak heights of both the predator and prey populations, which turns out to be an artifact of the estimation in the code. The Euler approximation is a somewhat crude method for solving differential equations and as such will result in some error associated with any solution. To improve the accuracy of these estimations, the time step for each iteration could be decreased, which results in diminishment of this artifact. However, to calculate the solution for the same amount of time, the number of iterations must be increased (since time is calculated by multiplying the time step by number of iterations).

6   Incorporating Alternative Assumptions to the Lotka–Volterra Model

As mentioned above, the Lotka-Volterra equations are based on assumptions about the nature of the predator–prey interactions and the environment in which they occur. Arguments could be made about the validity (or lack thereof) of such assumptions. Therefore, it is useful to explore different possible models based alternative assumptions. For example, let's assume the following:

- Rather than having access to an infinite amount of space and resource, there is a carrying capacity, K, for the prey population.

- The predator population does not necessarily infinitely feed on the prey population. Rather, if there was an abundance of prey available, the predator population would become sated prior to eating all the prey.

To incorporate these assumptions into the model, we can modify our two previous differential equations:

$$\frac{\mathrm{d}x}{\mathrm{d}t} = Ax\left(1 - \frac{x}{K}\right) - y \times Bf(x). \tag{5}$$

$$\frac{\mathrm{d}y}{\mathrm{d}t} = y \times Cf(x) - Dy. \tag{6}$$

where

- K is carrying capacity of the environment for the prey population. Notice how when x is small, the growth rate of prey approaches its maximum value A and as x approaches K (the total number of prey gets closer to its carrying capacity), the population will grow at a near-zero rate (indicative of competition within the prey population for food).

- $Bf(x)$ is the prey-consumption rate per predator density, which is represented by,

$$Bf(x) = \frac{Bx}{\alpha + x}. \tag{7}$$

where B is the maximum prey-consumption rate per predator density (per time unit per predator density), α is the prey population density at which half maximal prey-consumption rate per predator density is achieved. Notice when x is large, $Bf(x)$ saturates at B; when x is small, $Bf(x)$ increases almost linearly with x.

Our code can be modified, as shown below, to incorporate these changes (Fig. 12):

Note that in the above implementation we have updated some of the initial parameters, added the K and α parameters, and are now calling a new function *Predation_Num* outside our loop.

Running the above code gives the following result (Fig. 13):

Notice that the prey population oscillations never exceed the initial population density. Because the prey population is limited, the predator prey is also limited in how large it can grow. We could hypothesize that increasing the carrying capacity of the environment for the prey population could mitigate this effect (Fig. 14):

```
Lotka_Volterra_Expanded<-function(x_start=5,y_start=2,A=1.3,B=0.5,C=1.6,D=0.7,iterations=10000,time_step=0.01,alpha=1,K=3)
{
  x<-x_start
  y<-y_start
  dt<-time_step
  graph_frame<-c(0,x,y)
  for(i in 1:iterations)
  {

    dx<-(A*x*(1-x/K)-B*y*Predation_Num(alpha,x))*dt
    x_new<-x+dx

    dy=(C*y*Predation_Num(alpha,x)-D*y)*dt
    y_new<-y+dy

    x<-x_new
    y<-y_new
    new_data<-c(i*dt,x,y)
    graph_frame<-rbind(graph_frame,new_data)
  }
  matplot(x=graph_frame[,1],y=graph_frame[,c(2,3)],pch=20,xlab='Time',ylab='Population Density',
      main='Lotka-Volterra Simulaton',col=c('darkgreen','darkblue'))
  legend(x='topleft',legend=c('Prey','Predator'),col=c('darkgreen','darkblue'),lwd=5)
}

Predation_Num<-function(alpha,x)
{
  return(x/(alpha+x))
}
```

Fig. 12 Alternative Lotka–Volterra code

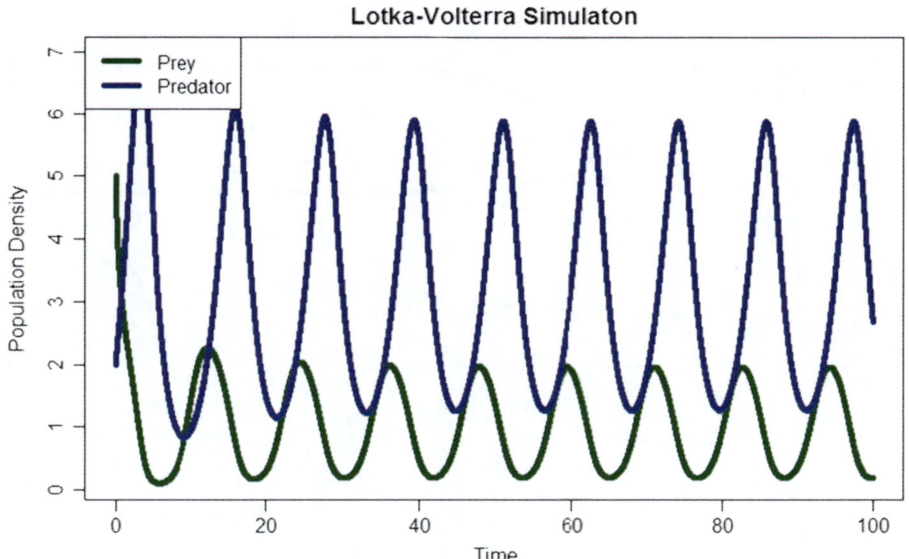

Fig. 13 Alternative Lotka–Volterra model result

```
Lotka_Volterra_Expanded<-function(x_start=5,
y_start=2,A=1.3,B=0.5,C=1.6,D=0.7,iterations=
10000,time_step=0.01,alpha=1, K=10)
```

Because the prey population is allowed to grow to a higher carrying capacity limit, the maxima of both populations become larger.

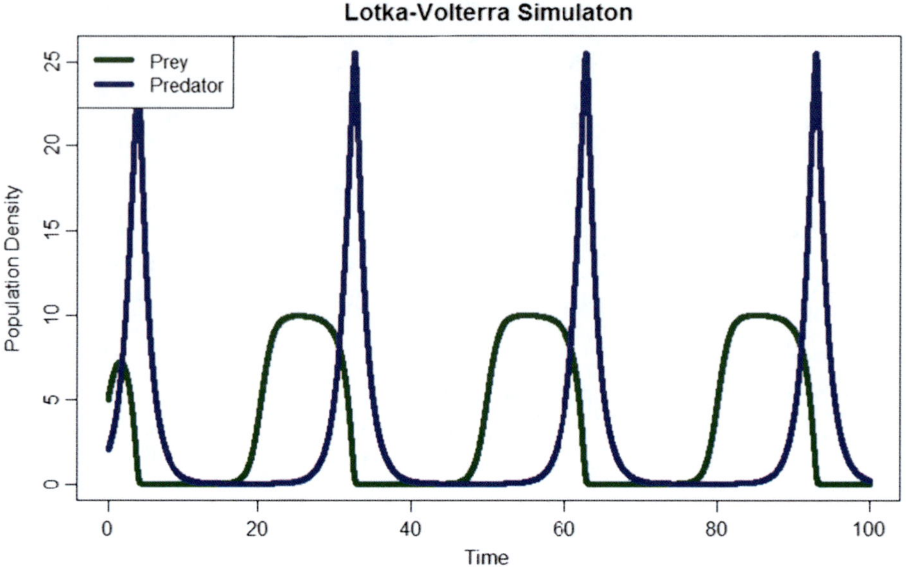

Fig. 14 Altered carrying capacity result

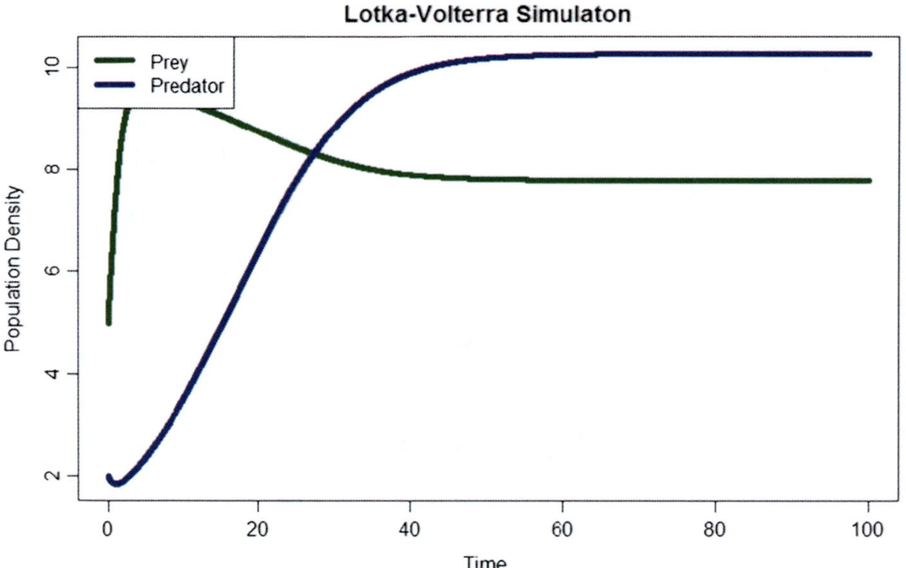

Fig. 15 Altered alpha value result

If we assume a larger α value (i.e., low predator affinity for prey from low capturing efficiency), a different pattern is likely to emerge. This can be tested as shown below (Fig. 15):

```
Lotka_Volterra_Expanded<-function(x_start=5,
y_start=2,A=1.3,B=0.5,C=1.6,D=0.7,iterations=
    10000,time_step=0.01,alpha=10, K=10)
```

Here we notice that the populations seem to reach a "steady state". This can also be demonstrated mathematically by setting the rate of change for each population equal to zero in Eqs. 5 and 6.

$$0 = Ax\left(1 - \frac{x}{K}\right) - y \times Bf(x). \qquad (8)$$

$$0 = y \times Cf(x) - Dy. \qquad (9)$$

From Eqs. 7 and 9, we obtain

$$f(x) = \frac{D}{C} = \frac{x}{a+x}. \qquad (10)$$

Thus, x^*, the steady state level of prey x is

$$x^* = \frac{D\alpha}{C-D} \sim 7.8. \qquad (11)$$

Combine Eqs. 8 and 11, y^*, the steady state level of predator y is

$$y^* = \frac{Ax^*\left(1 - \dfrac{x^*}{K}\right)}{Bf(x^*)} = \frac{Ax^*\left(1 - \dfrac{x^*}{K}\right)}{B\dfrac{x^*}{a+x^*}} = \frac{A}{B}(a+x^*)\left(1 - \frac{x^*}{K}\right) \sim 10.2. \qquad (12)$$

As shown above, when the rates of change for both populations are set to zero, the steady state population densities for both predator and prey correspond to the values shown in the above graph. At the steady state, the birth of prey equals consumption of prey, while the birth of predator due to consumption of prey equals the natural death of predator.

7   Using ODE Solver Libraries for Population Modeling

The above examples were pedagogical demonstrations of how R can be used to simulate population dynamics based on approximating differential equations with difference equations via Euler's method. There are other mathematical techniques for solving differential equations that are more accurate than Euler's method. Many of these techniques are incorporated into the R environment through *packages*, which can be thought of as an "add-on" to the current environment meant to serve a specific purpose. For example, one of the most popular packages for solving differential equations in R is the *deSolve* package [6]. This package is useful when solving *initial value problems*, which are differential equations where the initial values of the state variables (e.g., population densities of prey and predator) are given.

```
> install.packages('deSolve')
Installing package(s) into '/Library/Frameworks/R.framework/Versions/2.14/Resources/library'
(as 'lib' is unspecified)
trying URL 'http://cran.rstudio.com/bin/macosx/leopard/contrib/2.14/deSolve_1.10-3.tgz'
Content type 'application/x-gzip' length 3633054 bytes (3.5 Mb)
opened URL
==================================================
downloaded 3.5 Mb


The downloaded packages are in
 /var/folders/ks/s1d3k7qn2gb_kf4d3lnnhggr0000gn/T//RtmpDpLbjW/downloaded_packages
```

Fig. 16 Install package command output

We will illustrate how to use this package to solve the Lotka–Volterra equations (1) and (2). The first step is to install the *deSolve* package. This can be achieved using the *install.packages* command:

```
> install.packages('deSolve')
```

You should see something similar to Fig. 16 on your console.

Next, open a new R Script for editing and add the following to it (Fig. 17):

To make the code more understandable, let's break it down section by section. First, look at the following piece of code:

```
library(deSolve)

parameters<-c(A=1,B=0.2,C=0.08,D=0.5)
state<-c(x=5,y=2)
```

The *library* command is a way of telling the R environment to load the functionalities from the installed *deSolve* package for use in this script. Note that in this case we are storing the parameters of the Lotka–Volterra equations in a vector called *parameters*. The initial values of state variables x and y (prey and predator population densities, respectively) are stored in a vector called *state*. Next, we declare the actual function used for calculations:

```
LV<-function(t,state,parameters){
   with(as.list(c(state,parameters)),{
      dxOVERdt<-(A*x-B*x*y)
      dyOVERdt<-(C*y*x-D*y)
      return(list(c(dxOVERdt,dyOVERdt)))
})
}
```

Notice that d*xOVERdt* and d*yOVERdt* correspond to the dx/dt and dy/dt from Eqs. 1 and 2, respectively. The *with(as. list(c(…))* command is used to allow to access the values stored in the *parameters* (*A, B, C,* and *D)* and *state* (x and y) by their names

```
library(deSolve)

parameters<-c(A=1,B=0.2,C=.08,D=0.5)
state<-c(x=5,y=2)


LV<-function(t, state, parameters) {
  with(as.list(c(state, parameters)),{

      dxOVERdt<-(A*x-B*x*y)
      dyOVERdt<-(C*y*x-D*y)
      return(list(c(dxOVERdt,dyOVERdt)))


})
}


times <- seq(0, 100, by = 0.01)
out <- ode(y = state, times = times, func = LV, parms = parameters)


matplot(x=out[,1],y=out[,c(2,3)],pch=20,xlab='Time',ylab='Population Density',
        main='Lotka-Volterra Simulaton',col=c('darkgreen','darkblue'))


legend(x='topleft',legend=c('Prey','Predator'),col=c('darkgreen','darkblue'),lwd=5)
```

Fig. 17 ODE solver implementation

(this is a syntax step that must be taken for easily using the *deSolve* package). It is also important to note that we are returning a vector of the d*xOVERdt* and d*yOVERdt* values (the order is also important—rates of change for state variables must be returned in the same order they were listed in the *state* vector). This function will be applied iteratively by the *deSolve* package for model calculations.

```
times<-seq(0, 100,by = 0.01)
out<-ode(y=state, times=times, func=LV,
parms=parameters)
```

The first command uses the *seq* function, which is used to create a time sequence data object. Essentially, this data object will be used to tell the *deSolve* package to sample from $t=0$ to $t=100$ every 0.01 time steps (just like our code written previously). The second command uses the *ode* (ordinary differential equations) function from the *deSolve* package and stores the output in the matrix *out*. The *ode* is the default function for standard initial value problems, but there are many more functionalities in the *deSolve* package (for more information, type

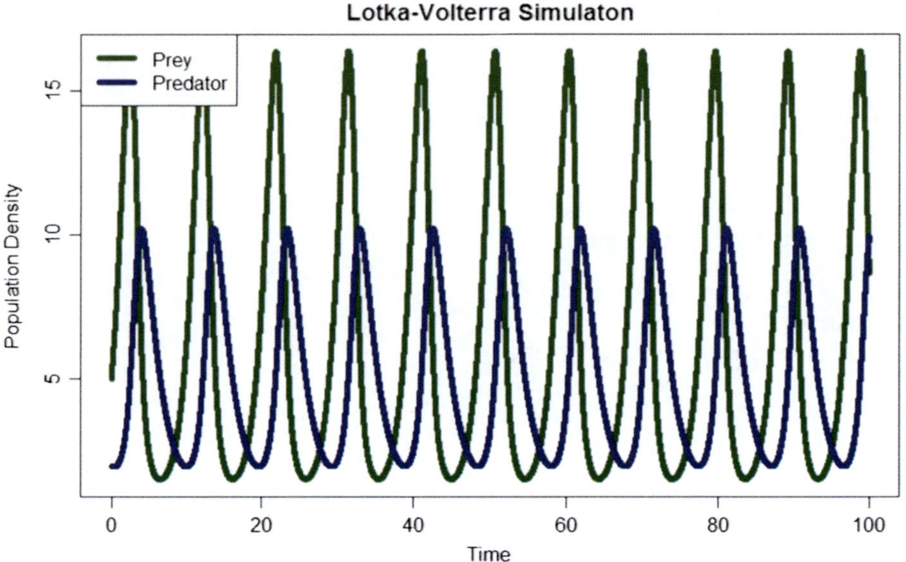

Fig. 18 Results using deSolve package

"?deSolve" in the Console). By default, *ode* uses an interface to an ODE solver written by Linda R. Petzold and Alan C. Hindermarsh [7]. However, there are other approaches, such as the Runge-Kutta method, that can be used by the *ode* function with the *method* parameter (for more information, type "?ode" into the Console).

Finally, because *out* is a type of matrix, it can be indexed and plotted like previous examples:

```
matplot(x=out[,1],y=out[,c(2,3)],pch=20,xlab='
Time',ylab='Population  Density',  main='Lotka-
Volterra   Simulation',col=c('darkgreen','darkb
lue'))

legend(x='topleft',legend=c('Prey','Predator'),
col=c('darkgreen','darkblue'),lwd=5)
```

Note that in this case, we haven't stored the bulk of the code in a function. Rather, every time the script is sourced, the entire code will run. Sourcing this file should give Fig. 18:

Notice how the peak heights of the predator and prey populations do not seem to increase, despite having a time step of 0.01. This can be attributed to the more accurate estimation in the *ode* function compared to the Euler's method.

Note: If, at any time, debugging is required, there are useful functionalities built into R that allows the user to halt execution of a function or script and examine the values stored in various parameters. One such function, browser(), can be placed at a given point in a script/function/loop and creates an interactive environment for the user to examine values. For more information, type "?browser" into your command console.

8 Conclusion

The purpose of this chapter was to demonstrate the power of using simple and open-source software like R to examine population dynamics represented by a mathematical model. With a mathematical model, hypotheses can be easily formulated and tested. The ability to represent abstract concepts in an intuitive manner can greatly facilitate the understanding of novel concepts and phenomena, leading to new insights in biological sciences.

Acknowledgements

The authors would like to thank members of the Shou lab (Björn F.C. Kafsack, David Skelding, Babak Momeni and Adam Waite), Sarah Holte, Jerry Davison, and Alex Hu for their critical feedback and insightful comments. Work in the W.S. group is supported by the W. M. Keck Foundation and the National Institutes of Health (Grant 1 DP2OD006498-01). R.G. is an NSF predoctoral fellow.

References

1. Malthus T (1798) An essay on the principle of population: an essay on the principle of population, as it affects the future improvement of society with remarks on the speculations of Mr. Godwin M, Condorcet and other writers. Electronic Scholarly Publishing, London, http://www.esp.org/books/malthus/population/malthus.pdf

2. Venables WN, Smith DM (2013) The R Core team. An introduction to R-Notes on R: a programming environment for data analysis and graphics version 3.0.1 (2013-05-16). http://www.cran.r-project.org/doc/manuals/R-intro.pdf

3. Lotka AJ (1925) Elements of physical biology. Williams & Wilkins, Baltimore, MD

4. Volterra V (1926) Variations and fluctuations of the number of individuals in animal species living together. J Cons Perm Int Ent Mer 3:3–51, Reprinted in R.N. Chapman, Animal Ecology, New York, 1931

5. Soetaert K, Petzoldt T, Setzer RW (2010) Solving differential equations in R. R J 2(2): 5–15

6. Soetaert K, Petzoldt T, Setzer S (2010) Solving differential equations in R: package deSolve. J Stat Softw 33(9):1–25

7. Hindmarsh A (1983) ODEPACK, a systematized collection of ODE solver (Stepleman R, et al. ed). IMACS Trans Sci C Comput 1:55–64

Chapter 16

Simulating Microbial Community Patterning Using *Biocellion*

Seunghwa Kang, Simon Kahan, and Babak Momeni

Abstract

Mathematical modeling and computer simulation are important tools for understanding complex interactions between cells and their biotic and abiotic environment: similarities and differences between modeled and observed behavior provide the basis for hypothesis formation. Momeni et al. (Elife 2:e00230, 2013) investigated pattern formation in communities of yeast strains engaging in different types of ecological interactions, comparing the predictions of mathematical modeling, and simulation to actual patterns observed in wet-lab experiments. However, simulations of millions of cells in a three-dimensional community are extremely time consuming. One simulation run in MATLAB may take a week or longer, inhibiting exploration of the vast space of parameter combinations and assumptions. Improving the speed, scale, and accuracy of such simulations facilitates hypothesis formation and expedites discovery. *Biocellion* is a high-performance software framework for accelerating discrete agent-based simulation of biological systems with millions to trillions of cells. Simulations of comparable scale and accuracy to those taking a week of computer time using MATLAB require just hours using *Biocellion* on a multicore workstation. *Biocellion* further accelerates large scale, high resolution simulations using cluster computers by partitioning the work to run on multiple compute nodes. *Biocellion* targets computational biologists who have mathematical modeling backgrounds and basic C++ programming skills. This chapter describes the necessary steps to adapt the original Momeni et al.'s model to the *Biocellion* framework as a case study.

Key words Discrete agent-based modeling, Partial differential equation, Adaptive mesh refinement, High-performance computing, Cell system simulation

1 Introduction

Discrete agent-based modeling maps a multicellular biological system to a collection of discrete agents. Discrete agent-based modeling has been widely used to model various biological systems [1, 3–5, 7]. Momeni et al. [5] studied spatial patterning in a community of yeast strains engaging in metabolic interactions through a combination of discrete agent-based mathematical modeling and wet-lab experiments.

Finding solutions using mathematical models often requires implementation of these models as a computer program.

Lianhong Sun and Wenying Shou (eds.), *Engineering and Analyzing Multicellular Systems: Methods and Protocols*,
Methods in Molecular Biology, vol. 1151, DOI 10.1007/978-1-4939-0554-6_16, © Springer Science+Business Media New York 2014

233

Producing a high-performance implementation that anticipates and accommodates easy revisions and refinements of a model as it evolves is a time-consuming task. Computational biologists often favor low programming effort over efficiency, flexibility, and even accuracy. Therefore, simulations tend to run much longer than performance-optimized code. Even incremental model updates can require significant code revision. For example, Momeni et al.'s model [5] partitions the simulation domain into fixed size cubic boxes and maps each cell to a single box. Once a cell divides, the daughter cell occupies one of the nearest neighboring boxes, instantly pushing surrounding cells outward to free up the space. This approximation of cell growth and shoving may be sufficient for the studied problem, but it may not work well for more complex problems—e.g., this approach cannot model a cell division that produces two cells differing in size.

Parallel computers ranging from multicore PCs to leadership class supercomputers provide significantly larger computing capacity than a single compute core. This computing capacity can address the computational challenges in simulating complex models when harnessed by *simulation software that serves widely varying multicellular biological system modeling requirements and runs efficiently on parallel computers*. However, implementing such software is difficult and time consuming. Even if computational biologists have access to cluster computers, without parallel computing, the advantage is limited to running multiple simulations in parallel—one simulation per compute core—resulting in slow turn-around time. In contrast, with efficient parallel software, a single simulation run can be partitioned across multiple computing cores, dramatically reducing the turn-around time.

Biocellion is a high-performance software framework that enables computational biologists without parallel computing expertise to exploit the power of parallel computers with only moderate programming effort. We briefly describe Momeni et al.'s model [5] and *Biocellion* [6] and illustrate the necessary steps to adapt Momeni et al.'s model to *Biocellion* as a case study.

1.1 Yeast Patterning Model Description

Momeni et al. [5] implemented a mathematical model with a range of parameters corresponding to different yeast strains engaging in various interactions. We consider only one instance in this article. Modification for other strains and interaction types is straightforward.

Two strains of yeast grow on top of a 24 mm thick agarose cylinder. One strain consumes lysine and also secretes adenine at a constant rate (say the *R* strain). The other strain consumes adenine and releases lysine on death (say the *G* strain). The two strains grow cooperatively and Momeni et al. demonstrated that strong cooperation promotes intermixing of the two strains in a three-dimensional community. We list model specifics necessary to adapt the model to *Biocellion*.

1. Yeast cells grow on top of a 24,000 μm thick agarose cylinder. Partial differential equations (PDE) are commonly used to model spatio-temporal variation of molecular concentrations in the extracellular space. We adopt this approach, and PDEs model adenine and lysine concentration changes in the agarose cylinder and the yeast cell community. Because the initial spatial distribution of the two yeast strains is uniform, a small fraction of the plate area in wet-lab experiment is already representative of patterns observed for the entire plate. Thus, only a fraction of the plate area in wet-lab experiments is considered in simulation. In Momeni et al.'s work [5], simulation domain height is set to include the entire agarose cylinder in the simulation area plus up to 300 μm above the agarose cylinder where yeast cells grow.

2. Periodic boundary conditions (both for PDEs and cell movements) are assumed in the x and y directions. A zero-flux boundary condition is applied at the bottom end of the agarose cylinder and at the top of the simulation domain—molecules and cells cannot cross the top and bottom planes of the simulation domain.

3. We map a cell to a sphere (instead of a fixed size cubic box in the original model) using *Biocellion*. The maximum cell diameter is 5 μm. The maximum cell volume is $\dfrac{4 \times \pi \times 2.5^3}{3}\ \mu m^3$. Yeast cells push against other cells when they are packed together.

4. We increase the volume of a sphere to model cell growth. The volume of an R strain cell after consuming a certain amount of lysine (say Δlysine) is $V_0 \times \left(1 + \dfrac{\Delta \text{lysine}}{\alpha_L}\right)$, where V_0 is the minimum cell volume (the volume of a cell right after cell division or one half of the maximum cell volume). α_L, the amount of lysine required to produce a daughter cell, is 2 fmol. Similarly, the volume of a G strain cell after consuming a certain amount of adenine (say Δadenine) is $V_0 \times \left(1 + \dfrac{\Delta \text{adenine}}{\alpha_A}\right)$. α_A, the amount of adenine required to produce a daughter cell, is 1 fmol. When a cell grows above the maximum cell volume, the cell divides into two equal-volume cells.

5. For an R strain cell, lysine uptake rate is $v_L = \dfrac{\alpha_L}{T_L} \times \dfrac{\phi_L}{K_L + \phi_L}$.

   For a G strain cell, adenine uptake rate is $v_A = \dfrac{\alpha_A}{T_A} \times \dfrac{\phi_A}{K_A + \phi_A}$.

The minimal population doubling times for lysine-requiring and adenine-requiring cells (T_L and T_A, respectively) are 1.76 and 1.98 h, respectively. The *Monod's constant* K_L, i.e., the concentration of lysine at which lysine-requiring cells grow at their half maximal growth rate is 1 μM. The Monad's constant K_A for adenine requiring cells is 0.1 μM. ϕ_L and ϕ_A are lysine and adenine concentrations in the extracellular space, respectively. An *R* strain cell secretes 0.08 fmol of adenine per hour, and a *G* strain cell releases 12 fmol of lysine on death.

6. The death rates are 0.054 h^{-1} and 0.018 h^{-1} for an *R* strain cell and a *G* strain cell, respectively.

7. Cells are randomly distributed on the agarose cylinder surface at the beginning of the simulation. Initial cell volume is set to a random value between one half of the maximum cell volume and the maximum cell volume to represent the range between a new-born daughter and a fully grown cell. The initial cell density is 500 cells per mm^2.

8. Initial lysine and adenine concentrations are set to zero.

9. Diffusion coefficients are 300 μm^2/s in the agarose cylinder and 20 μm^2/s inside a yeast colony for both lysine and adenine according to experimental measurements. Diffusion coefficients for grid boxes containing yeast cells are scaled down based on the total volume of the cells in a grid box—a grid box with low cell volume (a box in the colony-air boundary) has smaller diffusion coefficients than a box with high cell volume as implemented in the original model.

1.2 Biocellion Overview

Biocellion's design goal is to accelerate a wide range of discrete agent-based mathematical models of multicellular biological systems. This is challenging, because mathematical models of biological systems vary significantly. *Biocellion's* approach is to separate model specifics from common computational and parallel programming challenges. *Biocellion* asks users to provide model specifics, and *Biocellion* handles the remaining computational and programming challenges. Model specifics are expressed by the developer through modification of a library of C++ functions that comprise *Biocellion's* Application Programming Interface (API). The model library links to the *Biocellion* core framework at runtime. *Biocellion* output files can be visualized using Paraview (http://www.paraview.org).

Biocellion has three computational modules to simulate (1) individual discrete agent behavior, (2) direct physico-mechanical interactions between discrete agents, and (3) changes in the extracellular environment. *Biocellion* imposes a grid on the simulation domain to represent the state of the extracellular environment. In Momeni et al.'s model [5], cells reside on top of the 24 mm thick agarose cylinder, and the region where yeast cells grow is a small fraction of the entire simulation domain. The agarose cylinder is relevant only in tracking molecular concentrations in the model. Maintaining data structures for all three computational modules for the entire agarose cylinder can waste a significant amount of computing and memory. *Biocellion* imposes two different types of grids to different parts of the simulation domain to avoid such waste. *Biocellion* imposes an interface grid on a region where all three computational modules are executed, and computational modules communicate through this interface grid. *Biocellion* imposes a coarser PDE buffer grid on the region relevant only in solving PDEs—e.g., tracking nutrient concentrations in the agarose cylinder.

Biocellion decomposes the simulation domain into multiple partitions—users set the partition size. Users can impose either an interface grid or a PDE buffer grid for each partition. When users wish to run *Biocellion* on a cluster computer with multiple compute nodes (each node has multiple compute cores), *Biocellion* creates multiple compute processes, and every compute process works on a different set of partitions; note that a single process can exploit multiple compute cores in a single compute node to work on a single partition. Separate output files are created for different partitions.

Biocellion supports adaptive mesh refinement (AMR) to solve PDEs. AMR generates multiple levels of grids with different grid spacings based on the spatial resolution requirements of different simulation domain subregions. *Biocellion* asks users to set the number of AMR levels and the refinement ratio between two consecutive AMR levels—if the refinement ratio is set to 4, the coarser level grid spacing is four times larger than the finer level grid spacing. The finest grid spacing coincides with the interface grid spacing— *Biocellion* users set the interface grid spacing. The PDE buffer grid spacing equals the coarsest grid spacing in the AMR hierarchy. Users tag interface grid boxes with the desired AMR level. PDE buffer grid boxes are automatically tagged with the coarsest AMR level. *Biocellion* generates an AMR hierarchy (which is used to solve PDEs) based on this information. Note that the generated AMR hierarchy can have more fine boxes than the user input to improve efficiency (processing a large number of small boxes is inefficient) and guarantee correctness (coarsening a fine grid first and refining the coarsened fine grid should produce the original fine grid, or see the proper nesting condition in ref. [2]).

Figure 1 depicts the simulation domain (left) and the generated AMR hierarchy (right) in our experiment assuming 40 µm grid spacing, two AMR levels, and the refinement ratio of 4—we simulate with 5, 20, and 40 µm grid spacings.

Multicellular biological system simulation combines multiple biological processes such as cell movement, diffusion of molecules, and cell metabolic rate change. Different biological processes have

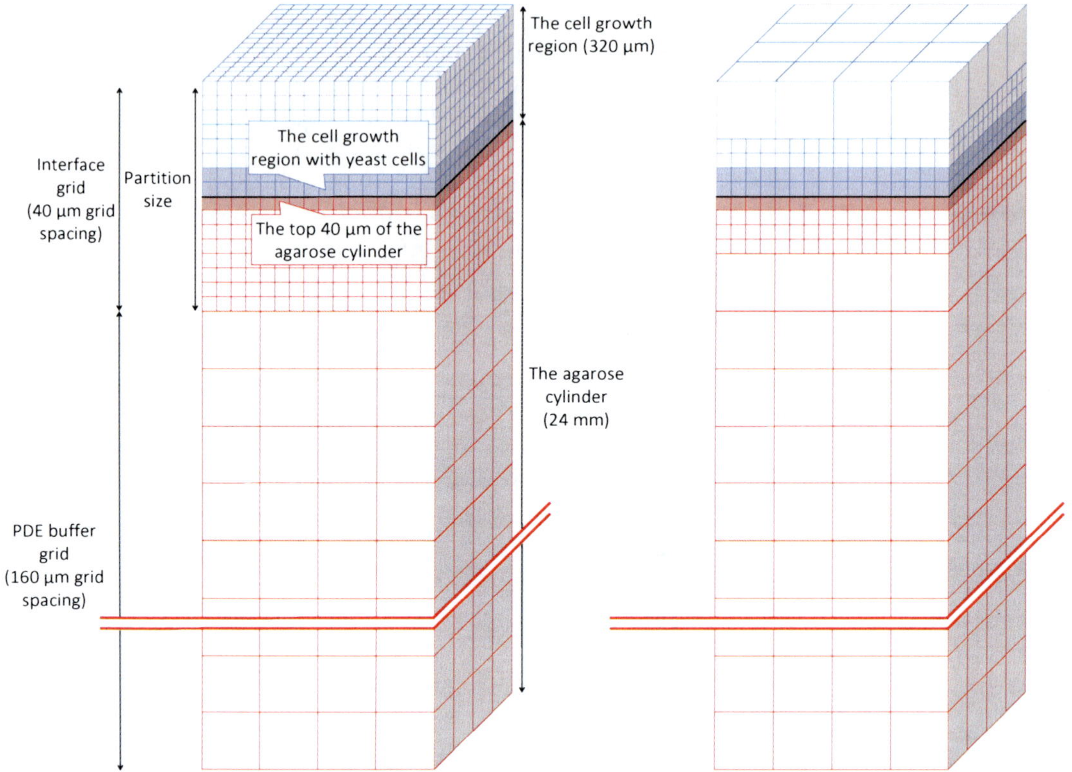

Fig. 1 *Biocellion* imposes two different types of grids to the simulation domain (*left*). An interface grid is imposed on the partition covering the cell growth region to simulate both cells and the environment. The remaining partitions in the agarose cylinder region are set as PDE buffer—these partitions are relevant only in solving PDEs to track lysine and adenine concentrations. Users tag interface gird boxes with the desired AMR level. We tag the interface grid boxes containing yeast cells (the *light blue boxes*) and the interface grid boxes at the top 40 µm of the agarose cylinder (the *light red boxes*) with the finer AMR level. The remaining interface grid boxes are tagged with the coarser level. *Biocellion* generates an AMR hierarchy (*right*) based on this information. Note that the generated AMR hierarchy has more fine boxes than the user input to satisfy the proper nesting condition [7]

different time step requirements to simulate the processes with sufficient accuracy. For example, simulating diffusion of molecules (by solving PDEs) often requires a significantly smaller time step size than the time step size required to simulate cell movement. To accommodate multiple time step size requirements in multicellular biological system simulation, *Biocellion* uses multiple time step sizes to simulate different computational modules and to communicate across the modules. The *baseline time step* is the largest time step used to simulate direct physico–mechanical interactions and discrete agent birth, death, and movement. The module computing direct physico–mechanical interactions communicates with the other two modules once per baseline time step. Discrete agent states and the state of the extracellular environment affect each other. For example, cell metabolic rate change affects the production and consumption rates of extracellular molecules, and this drives molecular concentration changes in the extracellular space. These two modules can be coupled more tightly by splitting a single *baseline time step* into multiple *state-and-grid time steps*. Variables associated with the grid imposed on the extracellular space can be updated either by model specific rules or by solving PDEs. Users can update the variables by model specific rules at the beginning and at the end of each *state-and-grid time step*. A single *state-and-grid time step* can be further partitioned to smaller *PDE time steps* to advance PDEs.

1.3   Porting Overview

The original yeast patterning model [5] partitions the simulation domain on top of the agarose cylinder into a set of fixed size cubic boxes (a box width is 5 μm), and a cell takes a single box. If a cell divides, the new cell tries to occupy one of the nearest neighboring boxes in the same z plane, if there is an empty box within the *confinement neighborhood* of 5-cell radius. The existence of confinement neighborhood was observed experimentally. If there is no empty box within the confinement neighborhood in the same z plane, the new cell occupies the box right on top of the mother cell box, and all the other cell boxes on top of the mother cell box are pushed upward.

Using *Biocellion*, we represent each cell by a sphere (*Biocellion* allows users to map a discrete agent to a different shape). A sphere can be located anywhere in the simulation domain above the agarose cylinder, and its radius changes to model cell growth. When a cell grows just enough to overlap with another, the model immediately introduces a force to push all spheres apart, thus modeling cell shoving in packed regions.

Figure 2 shows that the concentration of adenine changes smoothly in the simulation domain except for the cell growth region and the top 40 μm part of the agarose cylinder—adenine concentration changes smoothly even just 40 μm below the agarose cylinder top surface. We set the partitions at the agarose cylinder region (except for the top 40 μm part) as PDE buffer. *Biocellion* supports AMR which applies different grid resolutions to different parts of the simulation domain. In generating an AMR hierarchy, the region occupied by yeast cells and the top 40 μm of the agarose cylinder are tagged with the finest grid spacing, which is equal to the interface grid spacing. A coarser grid is imposed on the air region and the bottom part of the agarose cylinder.

We set the *baseline time step size* to 30 s and split a single *baseline time step* to 30 *state-and-grid time steps* to tightly couple cell metabolic rate change and nutrient concentration change in the extracellular space. *PDE time step* sizes to advance PDEs updating lysine and adenine concentrations are set identical to the *state-and-grid time step* size.

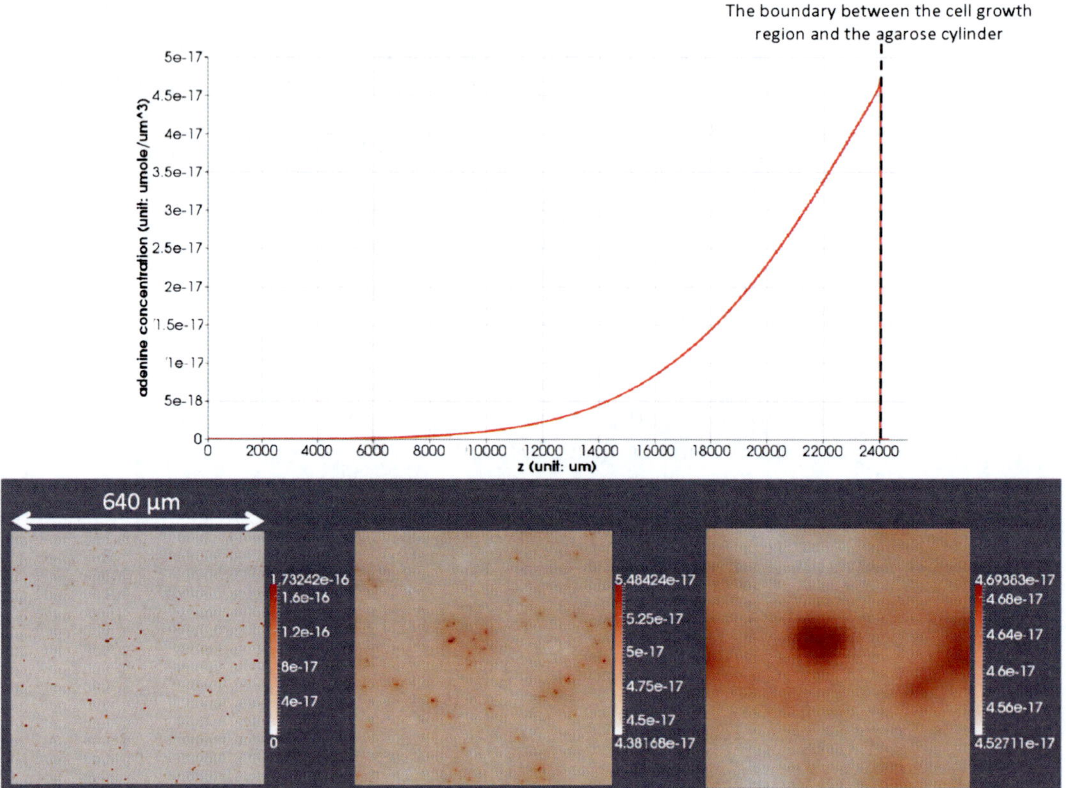

Fig. 2 Adenine concentrations in the simulation domain (unit: $\mu mol/\mu m^3$, $1~\mu M = 10^{-15}\mu mol/\mu m^3$). The *top figure* depicts the adenine concentration along the *z*-axis (passing the center of the simulation domain), with 0 and 24,000 being the bottom and the top of the agarose cylinder, respectively. The *bottom figures* show the adenine concentration at the *z* normal planes right on top of the agarose cylinder (*right*), right below the agarose cylinder top surface (*center*), and 40 μm below the agarose cylinder top surface (*right*), respectively. The adenine concentration has higher spatial variation near the agarose cylinder top surface. The spatial variation of the adenine concentration is significantly lower even just 40 μm below the top surface

We define four model specific variables (say rhs_{lysine}, $rhs_{adenine}$, $U_{scale,lysine}$, and $U_{scale,adenine}$) for each grid box in the interface grid. rhs_{lysine}, $rhs_{adenine}$ store the sum of the production and consumption rates of lysine and adenine, respectively. Lysine and adenine consumption rates are proportional to $\dfrac{\phi}{\phi + K}$, where ϕ is lysine or adenine concentration in the extracellular space and K (the concentration of metabolite at which half maximal consumption rate is achieved) is 1.0 and 0.1 μM for lysine and adeneine, respectively. $U_{scale,lysine}$ and $U_{scale,adenine}$ store the $\dfrac{\phi}{\phi + K}$ values for lysine and adenine, respectively. We want to limit the total amount of lysine or adenine consumed by cells in a grid box to be lower than the amount of lysine or adenine in the box plus an estimation of the amount of lysine or adenine diffuse into the box within a single *state-and-grid time step*. This prevents ϕ from becoming negative (especially when ϕ is small) without using a tiny time step size. This approach is accurate as long as our estimation of the diffusion rate is accurate. We use the explicit Euler method to estimate the amount of diffusion, which gives a reasonably accurate estimation for our choice of the *state-and-grid time step* size (1 s)—the molecular concentration gradient does not change significantly within 1 s. We scale $\dfrac{\phi}{\phi + K}$ to limit nutrient consumption. If the sum of rhs_{lysine} (or $rhs_{adenine}$) and the estimated diffusion rate multiplied by the *state-and-grid time step* size exceeds the lysine (or adenine) concentration of the box, we reduce the consumption rate and scale down $U_{scale,lysine}$ (or $U_{scale,adenine}$), so the net decrease of the lysine (or adenine) concentration based on the production, consumption, and estimated diffusion rates does not exceed the lysine concentration of the grid box. Model routines setting PDE parameters and model routines updating individual cell states can access these values to set nutrient uptake rates—this is necessary to assure that the total amount of nutrients consumed in solving PDEs coincide with the amount consumed by cells in updating individual cell states. We save rhs_{lysine} and $rhs_{adenine}$ to avoid computing the rates again when setting PDE parameters.

A *G* strain cell releases lysine on death. A relatively large amount of lysine is released in a short amount of time, and this forms a steep concentration gradient followed by a rapid gradient change due to diffusion. Accurately computing this transient gradient change requires a small time step size. However, cells react to concentration changes only gradually, so accurately computing the transient gradient change has little impact on simulation output. We have decided to spread released lysine to six neighboring boxes in the $\pm x$, y, and z directions to lower the initial concentration gradient. We implement this by updating the rhs_{lysine} variable of a neighboring grid boxes.

Figure 3 shows a simulation output. Simulation time is highly dependent on grid resolution and with 40 μm interface grid spacing (comparable to 50 *mum* grid spacing used in the Momeni et al.'s work [5]) to simulate 500 h of cell growth, a single simulation run takes 6.5 h on a workstation with a single 6 core microprocessor (Intel X5650 2.67 GHz). *Biocellion* also accelerates larger higher resolution simulations using multiple compute nodes.

2 Materials

Biocellion runs on multicore PCs, workstations, cluster computers, cloud computers, and supercomputers. This article pertains to multicore PCs. Running on different systems does not require model code changes. The current version of *Biocellion* runs only on ×86 compatible systems (PCs with Intel or AMD microprocessors are ×86 compatible). A 64-bit Linux operating system needs to be

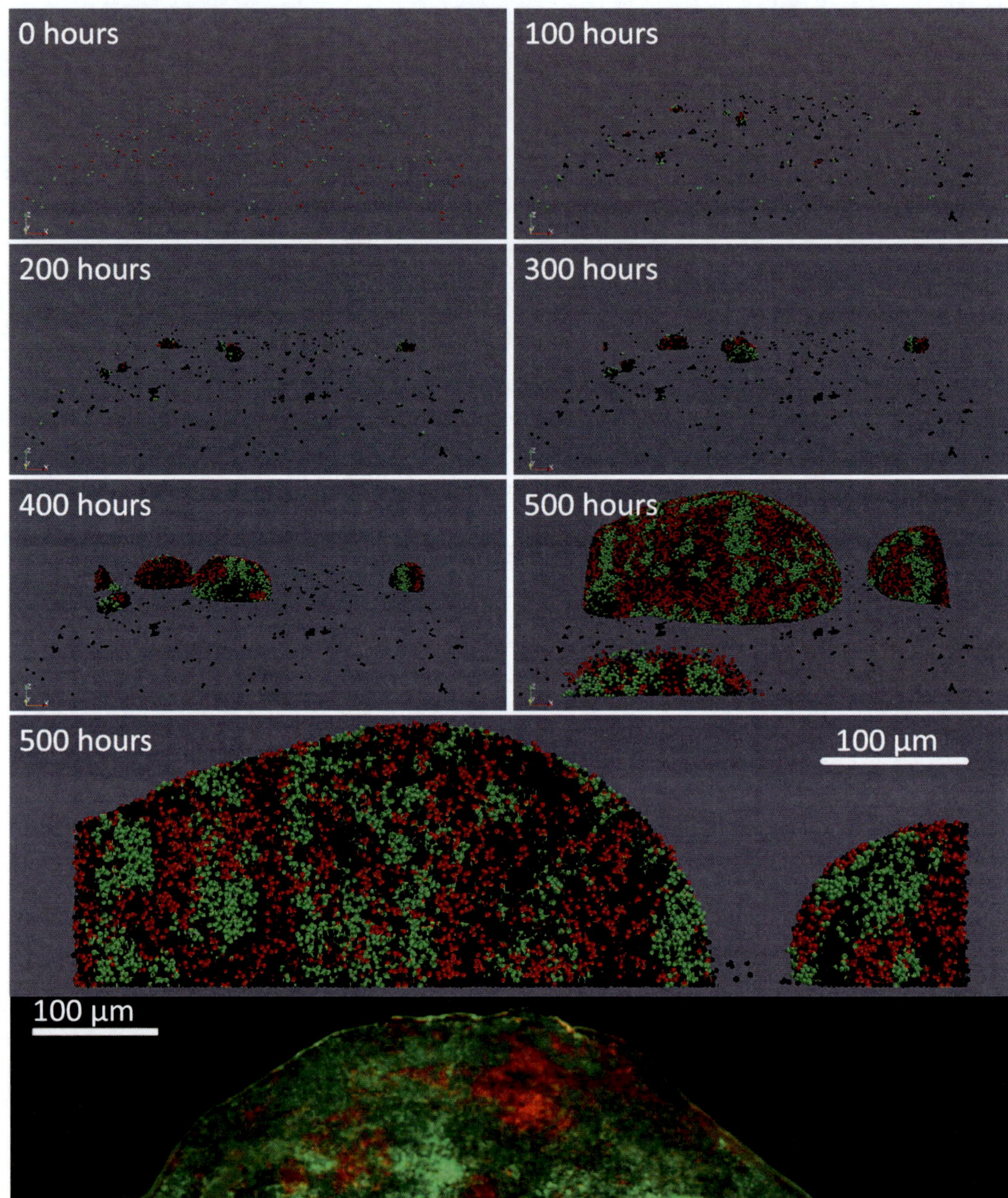

Fig. 3 Yeast cell growth (the top six figures) viewed from an oblique angle from the top, a 2D vertical cross-section of the yeast colony (*the next figure*), and a 2D vertical cross-section from the wet-lab experiment (*the bottom most figure*, reproduced from Momeni et al.'s paper [5]). *Red cells* are R strain cells, *green cells* are G strain cells, and *black cells* are dead cells

installed on the target system. Compiling *Biocellion* model code requires the GNU gcc compiler (pre-installed in most Linux systems), Intel icc compiler, or some other C++ compiler (we have tested only with gcc and icc). When compiling model code, users may set the check flag to verify their code or disable the check for higher performance (*See* **Note 1**). *Biocellion* requires the Intel Thread Building Blocks library, freely available from the thread building blocks homepage (http://threadingbuildingblocks.org).

2.1 Installing Intel Thread Building Blocks

1. Download the most recent stable version of Intel Thread Building Blocks library (version 4.2 or later is required) to the target system.

2. Unzip the downloaded tarball.

3. Update the `LD_LIBRARY_PATH` Linux environment variable to include the TBB library directory (in TBB 4.2, this is `$TBB_ROOT/lib/intel64/gcc4.1`).

2.2 Installing Biocellion

1. Unzip the *Biocellion* tarball.

2. Open Makefile.common under the *Biocellion* root directory.

3. Update `BIOCELLION_ROOT` to point to the *Biocellion* root directory.

4. Try "make" under the libmodel directory. This should compile the model library (`$BIOCELLION_ROOT/libmodel/interface/libmodel.so`).

3 Methods

Biocellion users provide model specifics by filling-in a set of C++ functions defined in five files under the `$BIOCELLION_ROOT/libmodel/model` directory: `model_routine_config.cpp`, `model_routine_agent.cpp`, `model_routine_mech_intrct.cpp`, `model_routine_grid.cpp`, and `model_routine_output.cpp`. These files include model routines to initialize the model, update discrete agent states, simulate direct physico–mechanical interactions, update the state of the extracellular space, and set simulation output, respectively. The entire model code for the Momeni et al.'s model [5] is available under `$BIOCELLION_ROOT/libmodel/model-yeast-patterning` for interested readers. Below, we present examples of how a model is specified.

3.1 Model Configuration

Model routines related to model configuration are defined in `model_routine_-config.cpp`.

1. *updateIfGridSpacing:* Set the interface grid spacing by filling the *updateIfGridSpacing* function body surrounded by/* MODEL START */and/* MODEL END */(Code 1). The interface grid spacing should be equal to or larger than the maximum direct physico-mechanical interaction distance. We map a cell to a sphere-shaped discrete agent and only consider cell shoving in evaluating short-range mechanical interactions.

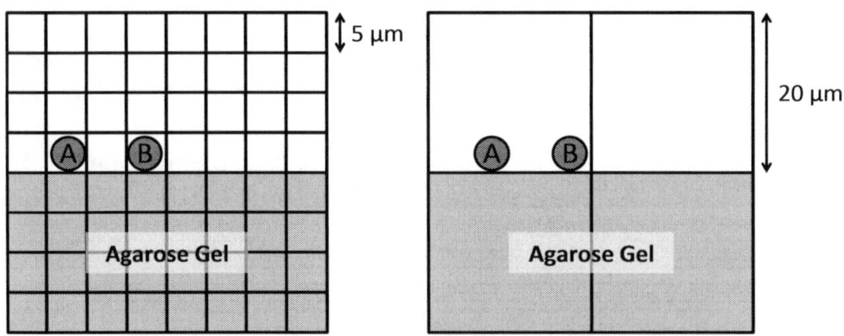

Fig. 4 Comparing 5 μm grid spacing and 20 μm grid spacing. Grid boxes outside the agarose cylinder with no cells have 0 diffusion coefficient. If cell *A* secretes a metabolite consumed by cell *B*, the secreted metabolite is delivered to cell *B* via diffusion through the agarose cylinder. However, if cell *A* and cell *B* are located in the same box (with 20 μm grid spacing), cell *B* can directly consume the molecules secreted by cell *A*

Two overlaping spheres are pushed apart to remove the overlap, and the maximum mechanical interaction distance cannot exceed the maximum cell diameter (5 μm). We start with the smallest interface grid spacing (5 μm) to minimize simulation artifacts (*see* Fig. 4 for an example). Larger values reduce execution time at potential loss of accuracy. Assuming fast diffusion, we may be able to adopt a larger grid spacing without significant loss of simulation accuracy. We also try 20 and 40 μm. Interested readers may experiment with different grid spacings to find the optimal grid spacing.

```
void        ModelRoutine::updateIfGridSpacing
(REAL& ifGridSpacing ) {
    /* MODEL START */
    ifGridSpacing = 5.0; /* 5.0, 20.0, or 40.0,
to set the interface grid
      spacing to 5.0, 20.0, or 40.0 um */
    /* MODEL END */
    return;
}
```

Code 1: Model routine to set the interface grid spacing.

2. *updateOptModelRoutineCallInfo*: Set the number of rounds to update variables associated with the interface grid at the beginning and at the end of a state-and-grid time step.

To set the model specific interface grid state variables properly (Subheading 1.3), *Biocellion* should be configured to invoke model routines updating interface grid state variables based on model specific rules once at the beginning of a *state-and-grid time step* and once at the end of the step (in order to reset the sum of lysine consumption and production rates, *see* **Note 2**).

3. *updateDomainBdryType*: Set domain boundary types. Periodic boundary conditions are applied in the x and y directions. A nonperiodic boundary condition is applied in the z direction, and cells are not allowed to pass the upper and lower end of the simulation domain in the z direction.

4. *updatePDEBufferBdryType*: Set the boundary type between an interface grid partition and a PDE buffer grid partition. The current version of *Biocellion* provides only one option (`PDE_BUFFER_BDRY_TYPE_HARD_WALL`), and discrete agents are not allowed to pass the boundary. This function is irrelevant to PDE boundary conditions.

5. *updateTimeStepInfo*: Update time step sizes. Set the *baseline time step size* to 30 s and split a single baseline time step to 30 *state-and-grid time steps*. Interested readers can experiment with different time step sizes.

6. *updateSyncMethod*: Update synchronization methods when a single variable is updated by multiple model routine calls (*see* **Note 2**). The only kind of short range cell–cell direct mechanical interaction we consider is cell–cell shoving, so the synchronization method for extra mechanical interactions is irrelevant. Set the synchronization method for grid variable updates to `SYNC_METHOD_DELTA` (*see* **Note 2**).

7. *updateSpAgentInfo*: Set discrete agent types. We consider three cell types (R and G strain cells and dead cells; dead cells do not grow and divide). A dead cell has one model specific variable storing the amount of lysine in the cell. This variable is used to calculate the amount of lysine to be released into the extracellular space.

8. *updatePDEInfo*: Set the grid state variables for lysine and adenine concentrations that are updated by solving PDEs. An AMR scheme is used with three levels (5 μm grid spacing) or two levels (20 μm grid spacing or 40 μm grid spacing). The finest level has the grid spacing equal to the interface grid. The size of a single PDE time step is set identical to the size of a single *state-and-grid time step*. A zero-flux boundary condition is applied in the z direction. Boundary conditions in the x and y directions are irrelevant, because periodic boundary conditions are imposed in *updateDomainBdryType*.

9. *updateIfGridModelVarInfo*: Set extra model specific variables associated with the interface grid. We add four variables per grid box (*see* Subheading 1.3).

10. *updateRNGInfo*: Set one random number generator to get random numbers with the uniform distribution.

11. *setPDEBuffer*: Set a partition as a PDE buffer partition if the top surface (in the z direction) of the partition is lower than the agarose cylinder height minus a small buffer region to

exclude the top part of the agarose cylinder (where molecular concentrations are highly localized). *setPDEBuffer* is called once per partition.

12. *setHabitable*: Set a grid box in the interface grid as a habitable box or an uninhabitable box. We set the boxes in the agarose cylinder to be uninhabitable, and cells are not allowed to move into an uninhabitable box.

3.2 Individual Agent Behavior

Model routines related to simulating individual agent behavior are defined in `model_routine_agent.cpp`.

1. *addSpAgents*: Randomly spread cells on top of the agarose cylinder to initialize the simulation. *addSpAgents* is called once per interface grid partition.

2. *updateSpAgentState*: Increase cell size based on the nutrient consumption rates. We use the scaled $\dfrac{\phi}{\phi + K}$ values (Subheading 1.3) to compute the nutrient consumption rates. This function is called once for every discrete agent in every *state-and-grid time step*.

3. *updateSpAgentBirthDeath*, *divideSpAgent*, and *adjustSpAgent*: Set whether a discrete agent will divide or disappear (*updateSpAgentBirthDeath*, see Code 2). *updateSpAgentBirthDeath* is called once for every discrete agent in every *baseline time step*. A cell divides if its size exceeds the maximum cell size. If a cell is set to divide, *divideSpAgent* is called. A cell divides into two cells in a random direction and the two resulting cells' volume is one half of the original cell's volume in the ported model. Dead cells remain in the simulation domain, so no discrete agent is set to disappear. If neither divide nor disappear is set, *adjustSpAgent* is called and this model routine updates the cell displacement based on the sum of the forces on the cell.

3.3 Physico–Mechanical Interaction Between Agents

Model routines related to simulating physico–mechanical interactions between discrete agents are defined in `model_routine_mech_intrct.cpp`.

1. *computeForceSpAgent*: Compute forces between pairs of discrete agents (*see* Code 3). This model routine is called once per cell pair that is within the maximum direct mechanical interaction distance, at every *baseline time step*. Force on two interacting cells is set based on the overlap between the two cells to remove the overlap by pushing the two cells apart.

```
void ModelRoutine::updateSpAgentBirthDeath(c
onst VIdx& vIdx, const SpAgent&
   spAgent, const AgentMechIntrctData& mechIn-
trctData, const Vector<NbrBox<
```

```
REAL> >& v_gridPhiNbrBox/*[elemIdx]
*/, const Vector<NbrBox<REAL> >&
   v_gridModelRealNbrBox/*[elemIdx]
*/, const Vector<NbrBox<S32> >&
   v_gridModelIntNbrBox/*[elemIdx]
*/, BOOL& divide, BOOL& disappear) {
   /* MODEL START */
   divide=false;
   disappear=false;
   if((spAgent.state.getType()==AGENT_TYPE_R_
CELL)||(spAgent.state.
   getType()==AGENT_TYPE_G_CELL))
{/* R or G straincells*/
   if (spAgent.state.getRadius
()>= MAX_CELL_RADIUS) {
      divide = true;
   }
  }
   else {/* dead cell */
   CHECK(spAgent.state.getType()== AGENT_
TYPE_D_CELL);
   }
   /* MODEL END */
   return;
  }
```

Code 2: Model routine to set whether a cell will divide, disappear, or neither divide nor disappear.

3.4 State Changes in the Extracellular Space

Model routines related to simulating state changes in the extracellular space are defined in `model_routine_grid.cpp`.

1. *initIfGridVar* and *initPDEBufferPhi*: Initialize grid state variables for the interface grid (*initIfGridVar*) and the PDE buffer grid (*initPDEBufferPhi*). *initIfGridVar* is called once per grid box in the interface grid, and *initPDEBufferPhi* is called once per grid box in the PDE buffer grid.

2. *updateIfGridVar*: Update interface grid state variables based on model specific rules (*see* Subheading 1.3). This function is called once per grid box in the interface grid.

```
void ModelRoutine::computeForceSpAgent(const
VIdx& vIdx0, const SpAgent&
   spAgent0, const VIdx& vIdx1, const SpAgent&
spAgent1, const VReal& dir/*
   unit direction vector from spAgent1 to spA-
gent0 */, const REAL& dist,
   VReal& force/* force on spAgent0 due to inter-
action with spAgent1 (force
```

```
    on spAgent1 due to interaction with spAgent0
has the same magnitude but
    the opposite direction), if force has the
same direction with dir, two
    cells push each other, if has the opposite
direction, two cells pull each
    other. */) {
    /* MODEL START */
    REAL R = spAgent0.state.getRadius () + spA-
gent1.state.getRadius ();
    REAL mag;/* + for repulsive force, - for
adhesive force */
    if (dist < = R) {/* shoving to remove the
overlap */
      mag = 0.5 * (R - dist);
    }
    else {/* adhesion */
      mag = 0.0;/* no adhesion */
    }
    for(S32 dim = 0; dim < DIMENSION; dim++) {
      force[dim] = mag * dir[dim];
    }
    /* MODEL END */
    return;
    }
```

Code 3: Model routine to compute force between two interacting discrete agents.

3. *updateIfGridKappa* and *updatePDEBufferKappa*: Set PDE parameter κ for the interface grid and the PDE buffer grid for each grid box. κ represents the cell volume exclusion in diffusion. As we are not considering cell volume exclusion, κ is set to 1.0 (0 % volume exclusion).

4. *updateIfGridAlpha* and *updatePDEBufferAlpha*: Set PDE parameter α. α sets the decay rate. We ignore lysine and adenine decay and set α to 0.0.

5. *updateIfGridBetaInIfRegion*, *updateIfGridBetaPDEBufferBdry*, *updateIfGridBetaDomainBdry*, *updatePDEBufferBetaInPDEBufferRegion*, and *updatePDEBufferBetaDomainBdry*: Set PDE parameter β. β sets the diffusion coefficient. β is set between two grid boxes sharing a face (*updateIfGridBetaInIfRegion* and *updatePDEBufferBetaInPDEBufferRegion*). Different model routines are called at the boundary between the interface grid and the PDE buffer grid (*updateIfGridBetaPDEBufferBdry*) and the simulation domain boundary (*updateIfGridBetaDomainBdry* and *updatePDEBufferBetaDomainBdry*). Code 4 sets the diffusion coefficient between two adjacent grid boxes in the interface

grid based on the yeast patterning model specifics by filling the function body of the predefined *Biocellion* model routine (*updateIfGridBetaInIfRegion*).

6. *updateIfGridRHSLinear* and *updatePDEBufferRHSLinear*. Set the PDE reaction term. *updateIfGridRHSLinear* sets the reaction term based on the lysine and adenine production and consumption rates for a grid box in the interface grid. No cells reside in the PDE buffer region and *updatePDEBufferRHSLinear* sets the reaction term to 0.0.

7. *updateIfGridAMRTags*. In solving PDEs, *Biocellion* users can apply different grid spacings for different regions; the finest grid spacing is the interface grid spacing. Users tag each box in the interface region with a desired AMR level. Boxes in the PDE buffer region are assumed to be tagged with the coarsest level. *Biocellion* (using CHOMBO [7]) generates an AMR hierarchy based on this information. We tag the boxes containing cells and the top 40 μm (in the z direction) in the agarose cylinder with the finest level. We tag the remaining boxes with the coarsest level.

3.5 Simulation Output

Model routines controlling simulation outputs are defined in `model_routine_output.cpp`.

1. *updateSpAgentOutput*. Color discrete agents. We color each discrete agent based on the cell type. We do not need to update extra output variables as we are mapping a discrete agent to a sphere. *See* Code 5 for our implementation for the *Biocellion* framework.

```
void ModelRoutine::updateIfGridBetaInIfRegion
(const S32 elemIdx, const S32
   dim, const VIdx& vIdx0, const VIdx& vIdx1,
const UBAgentData&
   ubAgentData0, const UBAgentData& ubAgent-
Data1, const Vector<REAL>&
   v_gridPhi0, const Vector<REAL>& v_gridPhi1,
const Vector<REAL>&
   v_gridModelReal0, const Vector<REAL>& v_
gridModelReal1, const Vector<S32
   >& v_gridModelInt0, const Vector<S32>& v_
gridModelInt1, REAL& gridBeta)
   {
/* MODEL START */
   REAL   z0   =   ((REAL)vIdx0[2]   +   0.5)   *
IF_GRID_SPACING;
   REAL   z1   =   ((REAL)vIdx1[2]   +   0.5)   *
IF_GRID_SPACING;
   REAL gridBeta0;
```

```
        REAL gridBeta1;
        if(z0 < AGAR_HEIGHT) {
        gridBeta0 = A_DIFFUSION_COEFF_AGAR[elemIdx];
         }
        else {
        REAL  scale  =  (REAL)ubAgentData0.v_spAgent.
size ()/(REAL)
          UB_FULL_CELL_CNT;
        if(scale > 1.0) {
          scale = 1.0;
         }
         gridBeta0 = A_DIFFUSION_COEFF_COLONY[elemIdx]
* scale;
         }
        if(z1 < AGAR_HEIGHT) {
        gridBeta1 = A_DIFFUSION_COEFF_AGAR[elemIdx];
        }
        else {
        REAL  scale  =  (REAL) ubAgentData1.v_spAgent.
size ()/(REAL)
          UB_FULL_CELL_CNT;
        if(scale > 1.0) {
        scale = 1.0;
         }
         gridBeta1 = A_DIFFUSION_COEFF_COLONY[elemIdx]
* scale;
         }
        if((gridBeta0 > 0.0) && (gridBeta1 > 0.0)) {
        gridBeta = 1.0/((1.0/gridBeta0 + 1.0/grid-
Beta1) * 0.5);/*
           harmonic mean */
         }
        else {
          gridBeta = 0.0;
         }
        /* MODEL END */
        return;
         }
```

Code 4: Model routine to set diffusion coefficient between two adjacent grid boxes in the interface grid.

```
    void ModelRoutine::updateSpAgentOutput(const
VIdx& vIdx, const SpAgent&
    spAgent,    REAL&    color,    Vector<REAL>&
v_extra) {
     /* MODEL START */
     color = spAgent.state.getType ();
     CHECK (v_extra.size () == 0);
```

```
/* MODEL END */
return;
}
```

Code 5: Model routine to set the discrete agent color variable (for visualization).

3.6 Setup for a Simulation Instance

Biocellion asks users to provide specifics of a simulation instance (e.g., simulation domain size, output directory) in an xml file. See `$BIOCELLION_ROOT/framework/main/yeast-patterning-5um.xml`, `$BIOCELLION_ROOT/framework/main/yeast-patterning-20um.xml`, or `$BIOCELLION_ROOT/framework/main/yeast-patterning-40um.xml` for examples (for 5, 20, or 40 μm grid spacing, respectively).

We set the required parameters first.

1. Set the number of base line steps to execute. We set this number to 60,000 (500 h, <*time_step num_baseline = "60000"*/>).

2. Set the simulation domain size (<*domain x = "128" y = "128" z = "4864"*/> in case we adopt 5 μm interface grid spacing). As we are using three AMR levels with the refinement ratio of 4 (with 5 μm interface grid spacing, refinement ratio is also set in this xml file), simulation domain size should be a multiple of 64. We set the domain size in the *x* and *y* directions slightly smaller than the size in [5], while setting the size in the *z* direction slightly larger than the size used in [5].

3. Set the simulation initialization method and the partition size. We set the initialization method to initialize within the code, and set the partition size to 64 (<*init_data partition_size = "64" src = "code"*/>). Alternatively, users can start from checkpoint data.

4. Set the output directory path, interval, file formats, and the number of extra output variables for each discrete agent (<*output path = "output_directory_path" interval = "120" particle = "pvtu" num_extra = "0" grid = "vtm"*/>).

We also set several optional parameters relevant to executing the yeast patterning model on multicore PCs.

1. Set standard output verbosity (from 0 to 5) to 1 (<*stdout verbosity = "1"*/>).

2. Set the number of threads to 12 for a multicore PC with 12 hardware threads (<*system num_node_groups = "1" num_nodes_per_group = "1" num_sockets_per_node = "1" max_load_imbalance = "1.2" num_thre ads = "12"*>). *num_node_groups*, *num_nodes_per_group*, *num_sockets_per_node* and *max_load_imbalance* are irrelevant for multicore PCs and are ignored.

3. Set the summary report (*Biocellion* allows users to print the summary of the interface grid variables, see the *Biocellion* user manual [6] for additional details), AMR regridding, and checkpoint intervals (*<interval summary = "10" load_balance = "120" regridding = "120" checkpoint = "600"/>*). *load_balance* is irrelevant to multicore PCs and is ignored.

4. Set the refinement ratio in applying AMR and other parameters controlling AMR hierarchy generation (*<amr refine_ratio = "4" fill_ratio = "0.5"/>*). See the *Biocellion* user manual [6] for additional details.

5. Set the optional parameters affecting the accuracy of the multigrid method used in solving PDEs (*<mg_parabolic_solve mg_num_pre = "3" mg_num_post = "3" mg_num_bottom = "3" mg_v_or_w = "v" mg_max_ite rations = "50" mg_epsilon = "-12" mg_hang = "-8" mg_norm_threshold = "-20"/>*). See the *Biocellion* user manual [6] for additional details.

---

4 Notes

1. *Biocellion* users can set the framework to perform checks on model routine outputs or input arguments of *Biocellion* utility functions called inside model routines by enabling *CHECK_FLAG = -DENABLE_-CHECK = 1* in `$BIOCELLION_ROOT/Makefile.model`. This often allows users to easily identify model program bugs such as using a random number generator without initialization or accessing a C++ STL vector array variable outside the array length. We recommend *Biocellion* users enable this option to verify their model routines prior to running full simulations. Once the model routines are verified, users should disable this check to expedite simulation.

2. Lysine released by a dying *G* strain cell is spread to seven grid boxes. The sum of lysine uptake and secretion rates for a grid box can be updated by seven different model routine calls (a model routine to update grid state variables is invoked once for every grid box in the interface grid in a single round). *Biocellion* asks users to set the synchronization method to properly update grid variables when a single variable is updated by multiple model routine calls. We set the synchronization method to `SYNC_METHOD_DELTA` to set the value by summing the differences from the initial value when updated by multiple model routine calls—e.g., if three different model routines set the value of variable rhs_{lysine} to 3, 5, and 9, respectively, then rhs_{lysine} is set to $3+5+9$ (assuming that the initial value is 0). We need to reset the sum of lysine uptake and secretion rates (rhs_{lysine}) to 0.0 to ensure that this scheme properly works. We configure *Biocellion* to invoke model routines to edit grid variables at the end of a *state-and-grid time step* to reset the variable to 0.0.

Acknowledgements

Support for this research was provided by the Extreme Scale Computing Initiative and the Fundamental and Computational Sciences Directorate, as part of the Laboratory Directed Research and Development Program at Pacific Northwest National Laboratory (PNNL). Portions of this work were conducted using PNNL Institutional Computing at PNNL. PNNL is operated by Battelle for DOE under contract DE-ACO5-76RLO 1830. B.M. is a Gordon and Betty Moore Foundation fellow of the Life Sciences Research Foundation.

References

1. Byrne H, Drasdo D (2009) Individual-based and continuum models of growing cell populations: a comparison. J Math Biol 58(4–5):657–687

2. Colella P, Graves DT, Johnson JN, Johansen HS, Keen ND, Ligocki TJ, Martin DF, McCorquodale PW, Modiano D, Schwartz PO, Sternberg TD, Van Straalen B (2012) Chombo software package for AMR applications design document. Lawrence Berkeley National Laboratory, Berkeley, CA

3. Ferrer J, Prats C, López D (2008) Individual-based modelling: an essential tool for microbiology. J Biol Phys 34(1–2):19–37

4. Galle J, Loeffler M, Drasdo D (2005) Modeling the effect of deregulated proliferation and apoptosis on the growth dynamics of epithelial cell populations in vitro. Biophys J 88:62–75

5. Momeni B, Brileya KA, Fields MW, Shou W (2013) Strong inter-population cooperation leads to partner intermixing in microbial communities. Elife 2:e00230

6. Pacific Northwest National Laboratory (2013) Biocellion 1.0 User Manual, 1.0 edition, Accessed Jul 2013

7. Xavier JB, Picioreanu C, van Loosdrecht MCM (2005) A framework for multidimensional modelling of activity and structure of multispecies biofilms. Environ Microbiol 7(8): 1085–1103

Index

Lianhong Sun and Wenying Shou (eds.), *Engineering and Analyzing Multicellular Systems: Methods and Protocols,*
Methods in Molecular Biology, vol. 1151, DOI 10.1007/978-1-4939-0554-6, © Springer Science+Business Media New York 2014